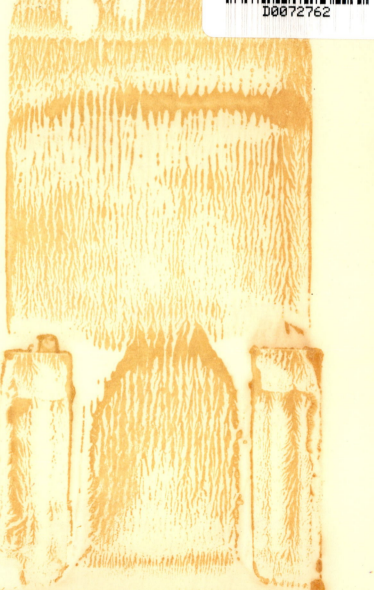

JET STREAMS

ELMAR R. REITER is a professor in the Department of Atmospheric Science at Colorado State University, Fort Collins, Colorado. Born in Wels, Austria in 1928, he attended the University of Innsbruck, receiving his Ph.D. in meteorology and geophysics in 1953 and the degree of "Dozent" in 1959. He served as research associate at the University of Chicago from 1954 to 1956.

In 1963 the University of Chicago Press published Dr. Reiter's handbook *Jet Stream Meteorology*, a translation from an earlier Austrian edition. Dr. Reiter has also written articles for scientific magazines and journals in several different countries.

Dr. Reiter has acted as scientific adviser to various branches of the aerospace industry, to NASA, and to the U. S. Army Missile Command. In 1963 he was scientific adviser to Project Topcat (Clear-Air Turbulence Measurement Program) in Australia, and has been director of several research contracts with U. S. Government agencies. He has lectured at scientific conferences in many European countries, the Soviet Union, Southeast Asia, and in Argentina.

ELMAR R. REITER

JET STREAMS

HOW DO THEY AFFECT
OUR WEATHER?

DOUBLEDAY & COMPANY, INC.
GARDEN CITY, NEW YORK

To BERNADETTE, REINHOLD, AND CHRISTA

THE SCIENCE STUDY SERIES

This book is one of a number that will appear in the Series through the collaboration of Educational Services Incorporated and the American Meteorological Society.

The Science Study Series was begun, in 1959, as a part of the Physical Science Study Committee's program to create a new physics course for American high schools. The Committee started its work in 1956, at the Massachusetts Institute of Technology, but subsequently became the nucleus of Educational Services Incorporated, of Watertown, Massachusetts, which has carried on the development of new curricula in several fields of education, both in the United States and abroad. The work in physics has had financial support from the National Science Foundation, the Ford Foundation, the Fund for the Advancement of Education, and the Alfred P. Sloan Foundation.

The purpose of the Series is to provide up-to-date, understandable, and authoritative reading in science for secondary school students and the lay public. The list of published and projected volumes covers many aspects of science and technology and also includes history and biography.

The Series is guided by a Board of Editors:
Bruce F. Kingsbury, Managing Editor
John H. Durston, General Editor
and Paul F. Brandwein, the Conservation Foundation and Harcourt, Brace & World, Inc.; Samuel A. Goudsmit, Brookhaven National Laboratory; Philippe LeCorbeiller, Harvard University; and Gerard Piel, *Scientific American.*

Selected Topics in the Atmospheric Sciences

The American Meteorological Society, with the objectives of disseminating knowledge of meteorology and advancing professional ideals, has sponsored a number of educational programs designed to stimulate interest in the atmospheric sciences. One such program, supported by the National Science Foundation, involves the development of a series of monographs for secondary school students and laymen, and since the intended audiences and the standards of excellence were similar, arrangements were made to include their volumes on meteorology in the Science Study Series.

This series within a series is guided by a Board of Editors consisting of James M. Austin, Massachusetts Institute of Technology; Richard A. Craig, Florida State University; Richard J. Reed, The University of Washington; and Verne N. Rockcastle, Cornell University. The Society solicits manuscripts on various topics in the atmospheric sciences by distinguished scientists and educators.

CONTENTS

Chapter I

THE DISCOVERY OF JET STREAMS

Winds

Violent or gentle, hot or cold, the winds puzzled and fascinated man long before he learned to record his thoughts on stone slabs or on papyrus. The ancient Greeks believed that the breath of deities caused the winds—the mild *Zephyrus,* god of the west wind, for instance. The Romans as they plowed the Mediterranean Sea with their galleys learned to dread the fury of the winds. But the daring Vikings used the winds to help them to probe the coastline of our continent. Although the masters of the Spanish galleons were adept at harnessing the power of the winds, the lashing gales and hurricanes were more than those proud ships could withstand, and the coral reefs of the Spanish Main and the Caribbean wrecked many a buccaneer's dream of early retirement among his treasures.

The early seafarers made quite an art of studying the various moods of the atmosphere. "Windjammer" captains knew about the winds that "blow trade" (that is, steadily) from the northeast toward the equator in our hemisphere, and from the southeast in the Southern Hemisphere. We still call these winds *trade winds.* They knew about the "doldrums" near the equator, where dead calms or light and fluctuating winds might shackle a sailing ship for days and days, while sailors sick with scurvy and low in fresh water and rum tried to curse the atmosphere into turning from a "dull-drum" into a "tantrum." And they were well acquainted with the horse latitudes,

near 30° North and 30° South, where light winds were
fickle as an old mare (Figure 1).

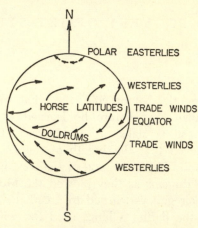

Fig. 1. The major wind systems man encounters in his daily
life at the earth's surface are shown on this map.

The famous men of those ages of conquest and ad-
venture were earthbound. Little did they care about
winds in the loftier regions of the atmosphere. But there
were some who dreamed of soaring like Icarus. Not all
were content with just watching the drift of clouds and
the flight of birds. Although Leonardo da Vinci's flying
machines never matured beyond the drawing board, the
idea, once born, did not die. In 1782, the hot-air balloons
of the brothers Montgolfier became a French fad. Pilâtre
de Rozier and d'Arlandes, the next year, actually took to
the air in one of these balloons. Berson and Süring, in
1901, reached the astounding height of 10,800 m (35,-
430 ft) in a gas-filled balloon. Auguste Piccard, later to
be known for his explorations of the ocean depths in the
bathyscaphe soared to 16,770 m (55,020 ft) in 1932, in
a balloon. In 1935, the Americans O. A. Anderson and

A. W. Stevens ascended to 22,256 *m* (73,347 *ft*), and in 1956 the U. S. Navy launched a manned balloon with a pressurized gondola which flew to 23,164 *m* (75,997 *ft*). Unmanned sounding balloons reached heights in excess of 35 *km* (114,000 *ft*).

From these flights it became apparent that winds and gales, unlike the gods of ancient Greece, are not earthbound. Clouds are not pulled or pushed by magic forces or by terrestrial forces; they drift with the air currents. The age of the Zeppelin or blimp, which we might compare with the age of the sailing ship, hastened exploration of the wind systems. But as the steamship pushed out the windjammer, so did Orville and Wilbur Wright doom the Zeppelin when, in 1903, they took their first hop in a propeller-driven contraption. Commercial airlines wove new arteries of trade and traffic through the atmosphere, and wind and weather became more and more important.

Enter: "The Jet Stream"

Aviation really came of age in World War II. The German Luftwaffe gave the first demonstration of effective airborne warfare. British and American bombers dared the hazards of bad weather. It seemed that the art of flying had become independent of the whims of the weather god: Takeoffs in poor visibility, instrument flying, and ground-controlled approach lost their daredevil flavor and became everyday routine. So, anyway, it may have appeared to the uninitiated.

In reality, however, the pilots made heavier and heavier demands on the weather forecasters, who had to use their imagination more than anything else if they were to give predictions for the data-void regions behind enemy lines. Whereas those magnificent men in their flying machines may have taken off on nothing more than a wing and a prayer, pilots of four-engine bombers de-

manded to know the winds en route, the clouds, visibility, icing conditions, and turbulence.

As long as the bombing runs and air battles kept close
to the ground, the forecasts proved to be more or less
satisfactory. Toward the end of the war, however, antiaircraft artillery and nasty little fighter planes had forced
the heavily laden and clumsy bombers, as well as the
poorly armed reconnaissance planes, to greater and
greater altitudes. Levels of six and seven kilometers were
outflown. And then, one day, it happened. With some
imagination we can reconstruct the events.

It was on Operation San Antonio,* the code name for
high-altitude precision bombing missions against the
Tokyo industrial compounds. High on the priority list of
targets was Nakajima's Musashino plant, at the edge of
a crowded suburb in the northwest part of Tokyo, some
ten miles from the Emperor's Palace. There a large part
of Japan's combat aircraft engines were manufactured.

Plans for San Antonio I called for daylight bombing
from 30,000 feet by ten to twelve squadrons, each of nine
to eleven planes and each plane carrying 5000 pounds
of bombs and 8070 gallons of gas. The B-29's of the XXI
Bomber Command from Isley Field, Saipan, were to fly
the missions. An F-13 reconnaissance plane had brought
back the information, on November 1, 1944, that at least
150 antiaircraft guns were distributed between Tokyo
and Funabashi, promising the bomber crews a flight
through hell over the target area. Furthermore, it was
estimated that around 400 to 500 fighter planes would be
ready to intercept the B-29's. Nevertheless, D-Day for
San Antonio I was set for November 17, 1944. The show
did not get off the ground.

"Most unusually and perversely, the prevailing easterly

* Information and quotes from *The Army Air Forces in World
War II*, Vol. V, edited by W. F. Craven and J. L. Cate. Chicago:
University of Chicago Press, 1953.

wind had veered around into the southwest, so that the customary take-off run would have to be reversed. At Isley this involved an uphill pull, hazardous at any time for combat-loaded B-29's, and suicidal in the steady rain which was falling as H-hour approached. A wing operations jeep notified airplane commanders that take-off would be delayed one hour, but after sixty minutes the rain showed no sign of abating. The mission was postponed for twenty-four hours."

So it went for a whole week—scheduling, postponing; scheduling, postponing. Nerves frayed to the breaking point, and combat crews griped that the B-29 was the best plane that never left the ground.

Finally November 24 dawned clear, and at 0615 the first plane, *Dauntless Dotty*, rolled down the strip. One hundred ten other B-29's followed, carrying 277.5 tons of bombs. The route took them west of the Bonin Islands, skirting a typhoon which was reported moving northeastward. The crews were briefed to attack along a west-to-east axis. Everything seemed to run like clockwork—at first.

On the way up to Tokyo mechanical failures forced six Superforts to turn back. These things can happen. The rest of the command droned on toward the target area, approaching it from the southwest. As they closed on the Japanese Islands, the winds picked up steadily. By the time they reached Tokyo, the formations flying at from 27,000 to 30,000 feet were swept into a 120-knot wind which gave them a ground speed of about 445 miles per hour. Below, an undercast almost completely obscured the target. Only twenty-four planes bombed the Musashino plant; sixty-four unloaded on dock and urban areas. Thirty-five aircraft had to bomb by radar. At such unexpectedly high wind speeds bomb drift was difficult to allow for, and the tangible results of the mission were discouraging. Low clouds hindered the F-13 reconnaissance planes. Their strike photos showed only sixteen bomb bursts in the target area. The Musashino management recorded forty-eight bombs (including three duds) in the factory area, with damage limited to 1 per cent of the building area and 2.4 per cent of the machinery. Casualties included fifty-seven killed and seventy-five injured.

San Antonio I was followed, on November 27, by San Antonio II, which did not fare much better. The "jet stream," almost as formidable a foe as AA guns and fighter planes, had made its entrance on the stage. The post-mortem report on the Precision Bombardment Campaign said:

"The most serious obstacle to successful bombardment was weather. Severe frontal conditions, frequently encountered on the trip north from the Marianas, increased fuel consumption, scattered formations, and made navigation so difficult that many crews missed the landfall entirely. Over the target crews rarely found atmospheric conditions suitable for precision bombing. The proportion of planes bombing visually had diminished progressively through the winter months: 45 per cent in December, 38 per cent in January, 19 per cent in February. Radar bombing seldom proved successful. The AN/ APQ-13 radar often malfunctioned at extreme altitudes, and radar operators were in general not sufficiently trained to get maximum results; even under skilled hands its accuracy was not up to the requirements of precision bombing. Cloud cover as a hindrance to bombardment was familiar to the AAF from bitter experience over Europe, *but the tremendous winds encountered at bombing altitudes over Japan offered a novel and most disconcerting problem.* [Emphasis mine.] With wind velocities reaching 200 knots and more, drift was difficult to correct and bomb runs had to be charted directly upwind or downwind. Attacking Japan's best-defended cities directly in the teeth of a 200-knot wind was unthinkable; going downwind the B-29's reached ground speeds in excess of 500 miles per hour, in which case neither bombsights nor bombardiers could function properly. Moreover, the high winds made it impossible for crews to make a second pass if the run-in failed; if a navigational error brought a plane in downwind from target it might not be able to attack at all."

The encounter with the jet stream was an important factor in forcing a change of tactics. On March 9, 1945, the stage was set for the first low-level incendiary raid on Tokyo. Three hundred thirty-four B-29's took off from Guam with about 2000 tons of bombs, reaching the target area after midnight, and flying as low as 4900 to 9200 feet. For 150 miles on the trip home the tail gunners could see the glow from the conflagration.

Even before American bombers encountered the mysterious winds over Japan, a lonely Nazi Junkers reconnaissance plane was cruising over the eastern Mediterranean, black crosses painted on its enormously long wings and fuselage. It was one of those drag-out flights with nothing to do but to watch the clouds and the British fleet far below. The plane was completely unarmed; its best weapon of defense was its enormous flight altitude of 17 km, which neither enemy aircraft or AA guns could reach. Expensive precision cameras made to order by the Reich's most skillful engineers were its only payload.

The crew had made this "milk run" many times before. They had passed Cyprus a while ago. It was time to head back home. The yards of film should keep Intelligence busy for a while. But that *dummkopf* weather observer—reporting a 170-knot wind before he lost his balloon from sight! He must have looked through a wine bottle instead of a theodolite.

A few hours later a scrambled "Mayday!" call reached the German base on Crete. "Strong headwinds—can't make it—have to ditch. . . ." The jet stream had claimed its victim.

After repeated reports of unbelievably strong winds over Japan, systematic research started on one of the most interesting phenomena the atmosphere has to offer. The jet streams now are well measured, well explored, and understood—at least in their large-scale aspects. Two decades after the discovery, we know that jet streams drastically affect our weather and climate. They carry heat and energy. They constitute one of the most important influences on the atmospheric circulation. But even nowadays they frustrate airlines and forecasters. Sometimes they carry aircraft at record-breaking speeds across continents and oceans, helping their flight with powerful tailwinds. Sometimes, however, extra fuel has to be carried at the expense of valuable payload in order to battle strong headwinds.

It was discovered that in winter and spring especially strong winds blow from west to southwest over Japan. On one occasion an aircraft over Tokyo measured winds of 650 *km/hr*. Wind reports of 400 *km/hr* (216 knots) are quite frequent for altitudes of between 10 and 14 *km* above sea level. Over the Mediterranean a similar current of air is found, although not as persistently as over Japan. It was, therefore, by no means a coincidence that the Allied and German air forces stumbled on this strange phenomenon at almost the same time. It was rather a matter of developing technology: How soon would man's flight climb to heights far beyond reach of even the most skillful birds?

It was discovered further that the strong winds occurred in relatively narrow bands, perhaps several hundred kilometers wide, but at the same time several thousand kilometers long. Because to the first pilots battling these winds it appeared as though the air had been blown out of a gigantic nozzle, the name "jet streams" seemed apt. It soon became apparent, however, that the comparison was not quite adequate. There is no nozzle in the sky from which these winds could be blown. They do not start at a specific place, as a water jet would start at the end of a garden hose. They are truly more like rivers of air, meandering about, accelerating and decelerating as they move along. But the name jet stream stuck nevertheless, like an advertisement slogan. With Americans and Germans stumbling on it at the same time, from the very instance of its discovery the jet stream became an international phenomenon. Naturally, this flashy name soon was misapplied to every little breeze that moved through the sky. Therefore, the World Meteorological Organization (WMO), with its headquarters in Geneva, Switzerland, felt itself compelled to put down some ground rules. The following definition was adopted:

"A jet stream is a strong narrow current, concentrated along a quasi-horizontal axis in the upper troposphere or in the stratosphere, characterized by strong vertical and lateral wind shears and featuring one or more velocity maxima."

The Structure of the Atmosphere

Some of the technical terms in the WMO definition may need explanation. We observe that in the lower atmosphere, temperature normally decreases with height up to a certain level. The *maximum* rate of this decrease is $1°$ Centigrade per 100 meters ($1°$ C/100 *m*) or $3°$ C per 1000 feet ($3°$ C/1000 *ft*). Under specific weather conditions the decrease may be considerably less. The layer over which such temperature decreases are observed is called the *troposphere* (Figure 2). Over the pole it is approximately 8 kilometers deep, over the equator approximately 16 *km*. It is in this deep atmospheric layer where most of our clouds and weather are observed.

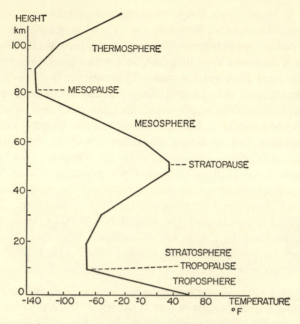

Fig. 2. The mean temperature structure of the atmosphere, plotted against altitude, shows the pauses as inflection points where the direction of the curve changes abruptly.

Above this layer the temperature remains almost constant at first, then increases up to a height of nearly 50 *km*. This layer is called *stratosphere*. The surface dividing the troposphere from the stratosphere is called the *tropopause*.

Above 50 *km* temperatures decrease again up to 80 *km*. This layer is called the *mesosphere. Stratopause* and *mesopause* are the names of the bottom and top surfaces of the mesosphere.

Beyond 80 *km* temperatures increase into outer space. They become much higher than 1000° C. This figure does not mean much, however, because there is hardly any air left to undergo these temperatures. Because of the lack of atmosphere, there is not much *heat* associated with these temperatures. A satellite sent to these altitudes is bothered very little. This lofty layer of increasing temperature is called the *thermosphere*.

What causes this peculiar temperature distribution in the atmosphere? Where we find high temperatures, there obviously must be a source of energy. This necessary relation makes an explanation very easy because the ultimate source of energy in the atmosphere is our sun.

The ground absorbs radiation. If you don't believe it, just walk barefoot across a paved road on a sunny day. Heat is given off to the atmosphere and carried upward by *thermals,* or convective bubbles. Sailplane pilots (and soaring birds) eagerly seek out these updrafts of warm air because the rising currents keep them aloft for a long time. As the air rises, it comes into the domain of lower pressure. (Pressure decreases as we go aloft because pressure is nothing but the total weight of the air column above. If we go up to greater height, there is less and less air above us; hence the decrease in pressure.) As the pressure of the rising air decreases, the air *parcel* cools, by the same principle on which your home refrigerator works. Now we know why the bottom of the troposphere (as shown in Figure 2) is warm: It is here that heat from the ground is received. And we also know why temperature decreases in the troposphere: The rising warm air will expand under the lower pressure of the environment and cool in this process of expansion.

If there is moisture in the air, it will condense into droplets as the cooling goes on. Clouds will form. Some types of clouds will show very clearly the shape and extent of rising bubbles of warm air. We call these *cumulus clouds*. If they become large and tall enough, they may give birth to showers and *thunderstorms*. Then we call them *cumulonimbus clouds*. In our latitudes these may reach altitudes well in excess of 30,000 *ft* (10 *km*). If you have ever observed a cumulonimbus cloud while it is forming, you will find that at first it resembles a monstrous cauliflower of brilliant whiteness in the sunshine (Plate 1). All of a sudden, in a matter of ten or fifteen minutes, its top becomes fuzzy. The cauliflower stops growing vertically and its top levels off. Apparently, the rise of the warm-air bubbles inside the cloud (which, incidentally, give the cloud its bubbly cauliflower appearance) has met with an obstacle: It has hit the *tropopause*.

From this and similar observations we may say that the tropopause poses a limit to rising air bubbles, or to *convective motions,* the technical language for these vertical currents. The stratosphere above is not bothered much by convective motions. Its temperature is controlled mainly by radiation and by horizontal air currents, such as the jet stream.

Thunderstorms quite frequently form near the jet stream, as we will see in Chapter VII. The action of the strong winds near the tropopause level may be seen when plumes of fuzzy cirrus clouds blow off and drift great distances away from the tops of the thunderstorm clouds (Plate 2).

The temperature peak at 50 *km* (Figure 2) and the increasing temperatures beyond 80 *km* are easily explained, too. Part of the solar radiation is absorbed as it passes through the atmosphere, which becomes denser and denser the farther the solar rays penetrate from outer space toward the ground. There is a delicate balance between the wave length, the amount of radiation absorbed, and the density of the atmosphere. The sparse molecules floating around at very high levels absorb X-rays and other very short wave radiation. This absorption causes the high temperatures of the thermosphere.

The radiation is so powerful that it knocks electrons out of atoms and molecules, which become electrically charged *ions*. We call these high levels of the atmosphere the *ionosphere*. These ionospheric layers reflect radio waves, and this reflecting property makes it possible for us to receive broadcasting programs over great distances (Figure 3).

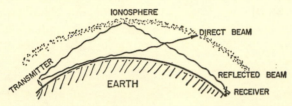

Fig. 3. Radio reception at a distance depends on the reflection of radio waves at the ionospheric layers. Direct beam from the transmitter (left) reaches few listeners, but reflected beam returns to earth to be picked up by receiver (right) far beyond the transmitter's horizon.

Ultraviolet radiation is absorbed near and below 50 *km*, mainly by the oxygen in the atmosphere—hence the temperature peak in that region. As we will see in Chapter VI, a strong jet stream—the so-called *polar night jet* —is associated with this "kink" in the vertical temperature profile. The energy of ultraviolet light knocks some of the oxygen molecules apart into two oxygen atoms. Some of these atoms recombine with other molecules of oxygen, forming ozone, or, in chemical notation, O_3. Ozone has a stinging odor and is highly toxic. Sun or ultraviolet lamps and electric sparks generate ozone in small quantities, causing the peculiar smell around some electric appliances.

From the foregoing we may see that the atmosphere acts as a filter against lethal doses of X-rays and ultraviolet light. Without this filter, life would have great difficulty maintaining itself on earth. The toxic quality of ultraviolet light even kills bacteria. That is why blue lamps are used to sterilize the pouring spouts of some soft drink machines.

As the atmosphere filters these short-wave components of radiation, it absorbs energy, which drives large circulation systems. Jet streams in the stratosphere, discussed in Chapter VI, feed on this energy reservoir provided by solar radiation.

Chapter II

WINDS AND WINGS

The dense air traffic crisscrossing continents and oceans nowadays requires accurate information on jet streams and winds in general. The reliability of forecasts issued by our meteorological offices depends to a large extent on the accuracy with which meteorological parameters, such as winds and temperatures, can be measured in the free atmosphere. If we can't trust the measurements, we can't trust the forecasts either.

It is fairly easy to measure winds near the ground: In first approximation, we need only stick a wet finger into the air. Or we may watch the behavior of grass and trees. Or we may judge the difficulty we have keeping on our feet while fighting our way through a gale. If more precision is required, we can measure winds with an anemometer—three or four little cups mounted on a revolving axis, whose spin is proportional to the speed of the wind.

Measuring winds in the upper atmosphere becomes a more complicated problem. We can watch the drift of clouds, but their speed may deceive us if we don't know exactly the height at which they are traveling. We need much greater precision than this.

Before we start to believe our information on jet streams, we had better take a look at the methods by which this information is obtained.

Measurement Units

In principle, it does not matter in which units we measure winds, be they miles per hour or "furlongs per fortnight." *To measure* means to compare the size or weight of some object with some *scale* that may be defined quite arbitrarily. Obviously, many people preferred scales of measurement which they could easily carry around. This is how the foot became a unit of length, with the inch (from the Latin *uncia*—one-twelfth) as a smaller unit. The difficulty arises from the whim of nature which equips some people with small feet, some with big. In order not to be shortchanged when shopping, those wearing small size shoes banded together against those with big feet, insisting that a *standard foot* be adopted as unit of length. With this arbitrary decision everyone was satisfied, although the convenience of carrying an accurate yardstick on one's body was lost.

In defining a length of time, nature quite conveniently offers the day as basic unit. As timekeepers became more accurate, it turned out, however, that days are not the same length throughout the year. Again, a *standard* or *mean solar* day had to be defined which is subdivided into twenty-four equally long hours (although some may seem longer than others).

We know that

$$\text{Velocity} = \frac{\text{distance traveled}}{\text{time}} \qquad (1)$$

or, in symbols,

$$V = \frac{L}{T} \qquad (2)$$

We may combine any two units of length and time to arrive at new units of velocity. Thus, we may conveniently express wind speeds in:

> feet per second (*fps*)
> (statute) miles per hour (*mph*)
> knots or nautical miles per hour (*kt*)
> meters per second (*mps*)
> kilometers per hour (*km/h*)

The latter two units belong to the so-called C.G.S. (centimeter, gram, second) system of measurement, which is one of the few beneficial bequests of the French Revolution. The meter is supposed to be the ten-millionth part of the distance from pole to equator. As it turned out, however, the measurement from which the length of one meter was derived was somewhat lacking in accuracy. Furthermore, the earth, with its mountains and valleys, is not a completely symmetrical sphere. Therefore, all meridians are not equally long, and even if the French had made their measurements correctly, we still would have to define a *standard meter*.

In order to find our way through the various units of speed customarily used in meteorology and other fields of physics and engineering, the following table provides handy conversion factors.

TABLE I

Conversion of Measurement Units

	fps	mps	One mph equals	kt	km/h
fps	1.0	3.281	1.467	1.689	0.911
mps	0.305	1.0	0.447	0.515	0.278
mph	0.683	2.237	1.0	1.152	0.621
kt	0.593	1.943	0.868	1.0	0.540
km/h	1.098	3.600	1.609	1.853	1.0

The units most commonly used for wind speed are knots and meters per second. In approximation, we may remember that 1 *mps* $\cong$ 2 knots. The table may also serve as a handy reference guide when arguing with a European traffic policeman about a speeding ticket.

What Is a Vector?

It is quite obvious that it does not suffice to indicate *speed* alone if we wish to describe a certain wind. It does make some difference whether we shiver in a 20-knot north wind or sweat in a 20-knot south wind. Thus we need two quantities to describe a wind completely:

its *speed* or magnitude
its *direction*.

Any quantities which have both magnitude *and* direction are called vectors (from the Latin word *vehere*, to carry). Other quantities which have a *magnitude* only are called *scalars* (from the Latin *scalae*, steps or ladder).

Dealing with scalars is simple: We just add or subtract them according to their sign and value. A typical scalar quantity is temperature. Suppose the thermometer shows 10° C. Then the temperature rises by 5° C. The final temperature reading will be 15° C.

Vectors are a little more tricky to handle. We already have met one of their representatives, *velocity*. It need not be wind velocity; it might be the velocity of a car, a train, an airplane, or a person. Let us assume that a small aircraft is heading straight eastward at a speed of 80 knots (Figure 4). A strong north wind of 60 knots carries

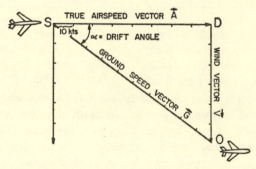

Fig. 4. Actual flight path from S to O of a plane traveling toward D at airspeed A but drifting with a crosswind V is shown in this vector diagram. The lengths of the sides of the triangle represent the relative speeds.

the plane off its course, and instead of arriving at point
D after one hour, the pilot finds himself over point O.
What was the actual speed of the airplane? Measuring
the distance from start S to point O, we find that it is not
80 + 60 nautical miles, but (from Pythagoras' theorem)
instead equals $\sqrt{80^2 + 60^2} = 100$ nautical miles. Since
the plane took one hour to cover this distance, it was
traveling at a speed of 100 knots, and the straight line
from S to O was the actual path of the aircraft.

In mathematical terms we have done the following.
We have taken the velocity vector of the aircraft (S to
D) and *added* to it the wind vector (D to O), arriving
at the actual displacement vector of the aircraft (S to
O). In order to compute the latter we had to perform a
vector addition, not the *arithmetic* addition we are accustomed to applying to scalar quantities.

Basic Navigation

Since we have been using an aircraft in our example of
vectors, we may as well adopt some of the professional
lingo of aeronautics.

The speed of the aircraft (in our example 80 knots)
is called *true air speed*. Since the aircraft swims in the
air, its air speed indicator will measure only the speed
of the aircraft relative to the surrounding air. The indicator is a so-called Pitot tube, an open tube pointing ahead
into the onrushing air stream. The higher the speed of
the aircraft through the air, the more pressure the air
will exert at the tube opening. This pressure is what is
measured to obtain the so-called *indicated air speed*.
Since the atmosphere becomes thinner as we go higher
up, less and less pressure will be measured by the Pitot
tube, even at the same air speed. Therefore, the indicated air speed will have to be corrected for this altitude

effect to yield the true air speed. This is the quantity we are interested in.

The vector of true air speed is made up of its magnitude (80 *kt* in our example) and the direction into which the nose of the aircraft points during flight. This direction is called the *true heading*. It is given in terms of compass degrees: north = 0 degrees, east = 90 degrees, south = 180 degrees, and west = 270 degrees. The indicator on board the aircraft may be a magnetic compass, showing the so-called *magnetic heading*. Because the magnetic poles do not coincide with the earth's North and South Poles, we will have to apply a correction to the magnetic heading in order to arrive at the *true heading* which we are interested in. These corrections may be obtained from aeronautical maps.

The wind in our example had a speed of 60 knots and a direction from the north. Wind directions are always given in terms of the angle *from* which the wind is blowing. Thus, a north wind has a direction of 0 degrees, an east wind has 90 degrees, etc. Speed and direction together make up the *wind vector* (D to O). The action of the wind causes the aircraft to drift off its original heading. The amount of this drift is given by the angle α, the so-called *drift angle*.

With this drift the aircraft will not follow the original route from S to D, but the *true course* or *track* from S to O. The actual speed at which the aircraft flies along this course we have already estimated: It was 100 knots. This we call the *ground speed,* because it is the speed at which the airplane travels over the ground. The *ground speed vector* is made up of the magnitude of the *ground speed,* and the *course angle*. The latter is given by the true heading (in our case, 90°, or east) plus the drift angle α.

We may now reformulate the statement we made at the end of the preceding section:

True airspeed vector + wind vector = ground speed vector

or

$$\vec{A} + \vec{V} = \vec{G} \qquad (3)$$

To indicate that we are dealing with vectors which have to be added by the special method, we draw little arrows above the appropriate quantities.

The Economy of Jet Streams

Now, with the vocabulary of a navigator, we may reconsider the discovery of jet streams in prosaic scientific terms. A B-29, let us say, flies at a true air speed of 250 knots, and at a true heading of 225° (toward the southwest). The plane struggles against a *headwind* of 190 knots from 225°. The drift angle will be zero. The ground speed, however, is only 60 knots. No wonder the unsuspecting navigator was baffled when such a fast plane (by World War II standards) almost stood still!

Another B-29, flying in the same jet stream, but at a heading of 45°, may experience *tailwinds* at full force. Again, the drift angle will be zero. The ground speed, however, is now 440 knots. This example shows how jet streams may be used to advantage for record-breaking long-distance flights. Since the jet streams within the altitude range of present aircraft have a preferred wind direction from the west, the advantage of tailwinds may be counted on only for eastbound flights. Pilots of westbound flights may want to make detours in order to avoid jet stream headwinds.

Such flight routes, deviating from the shortest distance between starting and landing point (the *great circle route*),* may take a full advantage of the day-to-day occurrence of jet streams. Even though a detour is made, the helping hand of a tailwind jet, or the avoidance of strong headwinds, will cause the flying time to be less than along the great circle. The *minimum flight path* technique actually seeks the route which requires the least flying time between two points, regardless of additional mileage required by detouring.

So, for instance, will the eastbound flight in Figure 5

Fig. 5. The longest way 'round may be shortest way home when an aircraft meets the jet stream. Though great circle route is shortest geometrical distance between A and B, pilot can cut air time by riding eastbound jet stream from A to B or detouring to avoid headwinds on westbound trip.

gain flying time by following the jet stream instead of pursuing the great circle route. The westbound flight will have the advantage of a short section of tailwinds if the pilot deviates to the north in the manner indicated.

The slower the aircraft, and the longer the non-stop flying distance, the more it pays to adjust the flight plan to the jet stream. Unfortunately, the increasing density of air traffic has put an end to the romantic postwar era when airline pilots hunted the jet streams. Now the Fed-

---

* This means that the greatest possible circle on a globe connects the two points. This circle has a radius equal to the earth's radius. It usually does *not* coincide with a straight line between the two points on a map.

eral Aviation Agency (FAA) sees to it that each airplane stays in its assigned air corridor—jet stream or no jet stream.

The jet stream, of course, is still there, but instead of outsmarting it by following or avoiding its course, we now have to outlast it by carrying extra fuel for any possible headwind condition. Fuel itself, fortunately, is not too expensive a commodity. At, let us say, 15 cents a gallon, an airline could easily afford to burn up a few extra barrels, even at the rate at which big jet engines swallow it. The trouble starts when we consider that instead of extra fuel the plane might have carried additional cargo, maybe at the rate of 8 cents per ounce of air mail.

Let us project ourselves to the desk of an airline president. We can easily calculate how the jet stream contributes to some of the hair graying on his executive temples. Assume that one of his 720's flies at 400 knots true airspeed, covering the distance between Washington, D.C. and Denver, Colorado, in roughly three hours, if there are no winds to contend with. With a 100-knot headwind all the way, the ground speed will be reduced to 300 knots, and the flying time will increase to four hours. Let's forget for the moment the annoyed passengers who missed their connecting flights and who have to be put up in hotels at the airline's expense. On a fuel consumption of approximately 12 *gals/min,* the plane will use 720 gallons more than it would have under zero wind conditions. The specific weight of kerosene is 0.82 *gm/cm³* or 6.85 *lbs/gal.* Thus, the extra amount of fuel weighs approximately 4940 pounds and costs $108.00 at 15 cents a gallon. (The actual price may be assumed less.) The loss in air mail postage, however, is $6323.20, figured at 8 cents per ounce. This is where the major loss comes in, even though we have to assume lesser rates for air freight.

We could have our company president shrug his shoulders, "Win a few—lose a few." On an eastbound flight with a tailwind jet stream and with fuel savings, the same plane should turn in a nice profit. Well, it is not quite so simple. If we figure that just as much mail travels east as west, and that the airline can handle only as many mail contracts as its *average* cargo space permits, we see the chances are that on the eastbound trip there may be empty space on the plane, held for the fuel that would have been taken had the jet stream not shown up in time. And fickle these jet streams are! They don't cancel their reservations in time if they don't plan to show up; they don't make a reservation long ahead of time if they plan to blow against the airplane's nose. Even a good 24-hour wind forecast will not help much when it comes to planning weeks ahead for a cargo shipment between industrial plants in the East and West.

Chapter III

HOW WE KNOW WHAT WE KNOW

Measurements of Wind with Theodolites and Pilot Balloons

Some devices measure wind direction only, such as a *wind vane*. Others, such as the *cup anemometer* (Plate III), measure wind speed only. The motion of the air exercises more pressure on the open face of each cup than on the convex side. The rate at which the anemometer spins may, therefore, be related to the wind speed. By using vanes and anemometers together, we can measure at any time the actual wind vector.

Anemometers and vanes usually need a fixed support. This mounting limits their use to measurements near the ground. How about measuring the *winds aloft* at heights far above the surface of the earth? The most widely used method makes use of *sounding* or *pilot* balloons. They are filled with helium gas, which is lighter than air. The balloons are sized to rise about 1000 feet per minute. When the wind blows, the balloon will not rise vertically but will drift off. If we keep track of the balloon position, say once every minute, we will be able to compute the mean wind within a layer which is 1000 feet thick (Figure 6).

Let us assume that at a given instant the sounding balloon is at point A somewhere up in the air. If there were no wind acting on the balloon, it would rise 1000 feet from A to B. This is the vector of the balloon *lift*. Due to the action of the *wind vector*, however, the balloon will after one minute be at point C. All we have to do is to find a way to measure the distance from B to C. This

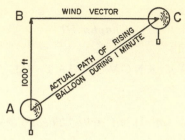

Fig. 6. Wind speed and direction can be read directly from vector diagram in which the hypotenuse of the triangle represents path of rising balloon.

distance would tell us immediately the mean wind speed in the 1000-foot layer from A to B or C in terms of *feet per minute* or *meters per minute*. After dividing by 60, we will have the wind speed in *mps*.

How can we measure this distance? We can use a telescope which permits accurate readings of angles, a so-called *theodolite*. We read the *elevation angle, φ*, every minute on the minute (Figure 7). Exactly one minute

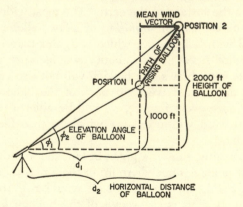

Fig. 7. Tracking sounding balloon with theodolite gives angles of elevation of balloon at Positions 1 and 2. Calculations with trigonometric functions of ϕ_1 and ϕ_2 give path of balloon and mean wind vector.

after the balloon has been released, it will be 1000 feet above the ground (Position 1). The distance d_1, by which the balloon has traveled horizontally in this minute, on account of the wind, may be calculated from trigonometry.* It is

$$d_1 = 1000 \; ft \cdot \cot \; \phi_1 \qquad (4)$$

d_1 in this case is given in feet.

Two minutes after release time, the balloon will be in Position 2, 2000 feet above the ground. The total horizontal distance which it has traveled is

$$d_2 = 2000 \; ft \cdot \cot \; \phi_2 \qquad (5)$$

During the first minute, the wind will have caused the balloon to drift horizontally by the distance d_1. Therefore, the mean wind speed in the first 1000-foot layer of the atmosphere is $d_1/60$, in *fps* if d_1 is measured in feet or in *mps* if d_1 is measured in meters.

The balloon's drift during the second minute is $(d_2 - d_1)$ and, therefore, the mean wind in the layer between 1000 and 2000 feet above ground is $(d_2 - d_1)$ / 60. In the layer between 2000 and 3000 feet, it will be $(d_3 - d_2)$ / 60, and so on, until we lose the balloon from our sight.

We still have one more problem to solve. So far we have assumed that the balloon keeps drifting in the same direction as it rises through the atmosphere. It will do so if the wind vectors in all the 1000-foot layers stacked on top of each other have the same direction; that is, if the wind does not turn with height.

What if the wind does shift? In addition to the elevation angle of the balloon above the level of the theodo-

---

* If you are not yet familiar with trigonometry and the use of trigonometric tables, you may use graph paper to work out the problems. If you are, but you have forgotten, the Appendix will refresh your memory.

lite, we measure the *azimuth angle*. This is the angle between a fixed geographic direction, say north, and the
direction of the balloon (Figure 8). The projection of

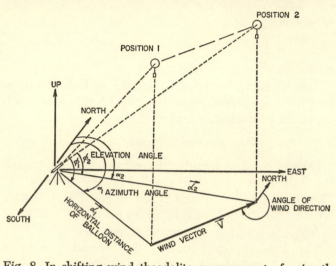

Fig. 8. In shifting wind theodolite measurement of azimuth
angles α_1 and α_2, in addition to angles of elevation, is required to calculate wind vector.

the balloon's position on the ground plane provides us
with the distance vectors $\vec{d_1}$, $\vec{d_2}$, etc. The difference
vector between $(\vec{d_2} - \vec{d_1})$, divided by 60, again gives
the wind vector in *fps* or *mps*.

If we wish to compute this problem in trigonometric
terms, it becomes more complicated. The change in azimuth angle between the two successive balloon positions
is $(\alpha_1 - \alpha_2)$. This subtraction gives us a triangle with
the two sides d_1 and d_2 and one angle $(\alpha_1 - \alpha_2)$.
We have to compute the direction and length of the
vector $\vec{V}$. In order to avoid lengthy computations we
may easily solve the problem graphically. We compute

the distances $\vec{d_1}$, $\vec{d_2}$, etc., for each minute of the balloon flight and plot them along the appropriate azimuth angle on a sheet of graph paper (Figure 9). We need only

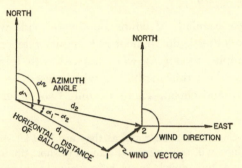

Fig. 9. Graphic solution to obtain wind vector requires horizontal plotting to scale of distances along appropriate azimuth angles. Connecting the end points of distance vectors so plotted gives wind vector. Angle of wind vector from north gives wind direction.

connect the end points of these vectors, 1, 2, 3, 4, 5, etc., to obtain the wind vectors. All we have to do is to measure the direction of each of these vectors, starting the angle from north (Figure 10) to get the *wind direction*.

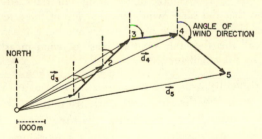

Fig. 10. Horizontal projection (by method of Fig. 9) of balloon's path gives a succession of wind vectors, each holding for a thousand-foot layer.

Then we measure the length of each wind vector with
the appropriate scale of our graph paper, and we have
the *wind speed.*

Radio-Theodolites

Early measurements of winds aloft were made with op-
tical theodolites; pilot balloons of known rate of ascent
(for example, 1000 *fps*) were tracked with telescopes.
Even nowadays this method is widely used because it
is effective and cheap. There are only two things wrong
with it:

> 1. If the winds are strong, as they are in a jet stream,
> the balloon will travel great distances in short times, and
> become a tiny speck even in a strongly magnifying
> theodolite. When the observer has lost sight of the bal-
> loon, he cannot measure the wind. This explains why jet
> streams were discovered so late, though pilot balloons
> had been in use for many years.
> 2. If there are clouds in the sky, the balloon may hide
> behind one of them, and again put an end to measure-
> ment. As we will see later, there frequently is bad
> weather associated with jet streams. If we can measure
> only in fair weather, and not in bad, our upper wind
> data will be strongly *biased.* We learn about winds at
> levels above the clouds only during fair days. If we
> assume that the same winds are blowing at these levels
> during cloudy days, we may be very wrong. Meteorolo-
> gists call this the "fair weather bias" of wind measure-
> ment by pilot balloon.

This difficulty was overcome during and after World
War II with the development and installation of radio
tracking devices, the so-called *radio-theodolites.* These
instruments are constructed on various working principles, some of them fairly simple. The Finnish radio-
theodolite, for instance, uses the radio signal emitted

from a transmitter carried aloft by the balloon. The signal also relays to the ground information on atmospheric pressure, temperature, and humidity (Plate IV), and thus serves a dual purpose. Such a device is called a *radiosonde* or, since it measures winds at the same time, a *rawinsonde*.

At the ground, a system of two antennas, one pointing in the north–south direction (A to B) and the other in the east–west direction (C to D) (Figure 11) receives

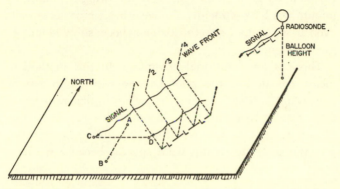

Fig. 11. Radiosonde on balloon sends radio wave signal to two perpendicular antennas, one pointing north–south, the other east–west. Differences in time of arrival of wave fronts at end points of antennas permit calculation of position of balloon.

the radiosonde signal. The electromagnetic waves of the radio signal will arrive at the antenna site in the form of plane wave fronts 1, 2, 3, 4, etc., spaced at the wave length L of this radio signal. Since radio waves travel very fast ($299{,}790$ km/sec), the direction from which these plane wave fronts arrive at the antennas gives the present direction of the balloon in the sky. We see in the diagram that the waves in the present example will first arrive in point *D* before they reach point *C*. They will also hit point *A* before they travel over point *B*. From these "microscopic" time differences measured in microseconds [1 microsecond (*msec*) equals $\frac{1}{1,000,000}$ *sec*] we are able to determine azimuth and elevation angle of the balloon. Knowing the balloon height, for instance, from its ascent rate (1000 *ft* per second) and measuring the time since its release, we can calculate its position in space. By following it minute by minute, we have a means of measuring winds accurately, even if we cannot see the balloon because of clouds or great distance.

There is an additional advantage to this method of measurement. For the pilot balloon we had to assume a constant rate of ascent. But what if this rate varies from minute to minute, for instance through the action of vertical up-and-down drafts? The answer is very simple. We compute wrong wind speeds, because we have

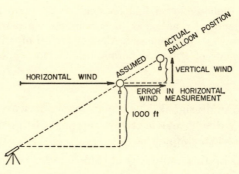

Fig. 12. Error in estimation of rate of ascent of balloon leads to the error shown (heavy horizontal line) in calculation of horizontal wind vector.

assumed the wrong balloon position; we placed the balloon where it would have appeared under the same elevation angle but at "normal" rate of ascent. An example
is given in Figure 12, showing how winds are under-
estimated for too great a rate of ascension.

If, however, a sounding gives information on pressure
and temperature, we may compute the *height* of the
balloon at any given moment without having to rely on
an *assumed* rate of ascent. Our calculations become
much more accurate and reliable. From the anomalous
variations in the ascent rate during one sounding run
we may even obtain an estimate on vertical wind velocities in the atmosphere, especially if they are rather large,
such as in cumulus clouds or in wave flow over mountains. Frequently we find that the winds form a series
of "waves" after the air crosses a mountain range. The
balloon will rise faster or slower depending on whether
it moves in the upward or downward branch of a wave
(Figure 13). Sometimes these waves are visible from

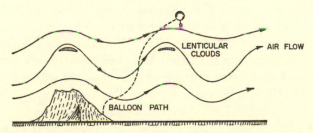

Fig. 13. Wave motions in the air, caused by flow over mountains, may account for strong fluctuation in the ascent rate of
balloon.

lens-shaped clouds (so-called "lenticular clouds") which
form in the crests of such waves (Plate V).

Radar

Even more advanced than radio-theodolites is *radar*. The acronym stands for "*ra*dio *d*irectioning *a*nd *r*anging." This name implies that the equipment also measures *range* or distance between antenna and target, in our case the balloon. A series of short radio pulses is sent out by the antenna. These pulses are reflected by the balloon and bounced back to the receiver. Since the speed of electromagnetic waves is known to be 299,790 *km* per second, the *delay time* between the emission and the reception of the signal is given by the equation

$$\text{Delay time } (T) = \frac{2 \times \text{Balloon Distance}}{\text{Speed of Waves}} \qquad (6)$$

From this equality we can easily compute the distance of the balloon from the antenna.

Suppose a balloon is 30 *km* away. There will be a lapse of $\dfrac{2 \times 30}{299,790}$ seconds, or approximately 200 microseconds, from the instant the signal is emitted by the radar set to the instant the reflected signal is received.† In order to reflect a sufficiently strong return signal, the balloon carries a *reflector* aloft. This may be a tetrahedron of aluminum foil (Plate VI). The balloon itself may be used as reflector if its upper side is coated with a thin metal foil. The concave shape of this metal half-sphere will provide good reflection of the radar signal.

Radar tracking differs slightly from theodolite tracking in the geometry used. The latter computes changes of the balloon's horizontal distance from changes in height, elevation angle, and azimuth angle. *Radar* measures

† Want to try one on your own? If a radar signal were reflected by the moon (distance from the earth approximately 380,000 *km*), what will be the delay time between outgoing and return signal?

changes in straight distance to the balloon, or in *slant range* as it is called, and azimuth angle (Figure 14).

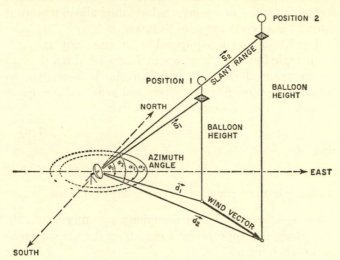

Fig. 14. Measuring the wind vector by tracking balloon with radar leads to this geometry from which d_1 and d_2 can be calculated by applying the Pythagorean Theorem.

Knowing the height of the balloon, we may compute the changes in horizontal distance (d), and from this the wind vector.

We may apply Pythagoras' theorem, and calculate

$$\text{Distance} = \sqrt{(\text{slant range})^2 - \text{height}^2} \qquad (7)$$

$$d_1 = \sqrt{S_1{}^2 - 1000^2} \qquad (8)$$

and

$$d_2 = \sqrt{S_2{}^2 - 2000^2} \qquad (9)$$
$$\text{etc.}$$

$\vec{d_1}$ and $\vec{d_2}$ may be plotted along the appropriate azimuth angle on a graph similar to the one shown in Figure 9 or 10. The connecting vector between the end points of the vector lines $\vec{d_1}$ and $\vec{d_2}$ gives the wind vector (in feet per minute) in the layer which the balloon traversed during the second minute of its ascent.

If our radar set is equipped to measure not only azimuth angle but also elevation angle, we need not depend on an assumed rate of ascent of the balloon, nor on height computed from pressure and temperature. From the rules of trigonometry‡ we know that

$$d_1 = s_1 \cdot \cos \phi_1 \qquad (10)$$

and

$$d_2 = s_2 \cdot \cos \phi_2 \qquad (11)$$

Again, we may proceed to calculate the wind vector (that is, wind speed and direction) by entering $\vec{d_1}$, $\vec{d_2}$, etc., along the appropriate azimuth angles, as shown in Figures 8, 9, 10, and 14.

Vertical Wind Profiles and Hodographs

Wind speeds measured from rawinsonde balloons may be entered in a diagram which has speed and height as coordinates. Quite frequently we find that the wind increases strongly with height up to a certain level. This is called the *Level of Maximum Wind (LMW)*. In the example of the wind sounding measured at Green Bay, Wisconsin, on April 19, 1963, 00 Greenwich Mean Time (GMT), this level is found at 10 *km* (Figure 15). Above this level the wind speed decreases again, at first slowly, then rapidly. A diagram, such as the one shown in Figure 15, is called a *vertical wind profile* because it exhibits the atmospheric motions in a profile plane.

Measurements like the ones shown are made twice a day at a large number of stations in both hemispheres.

‡ See Appendix.

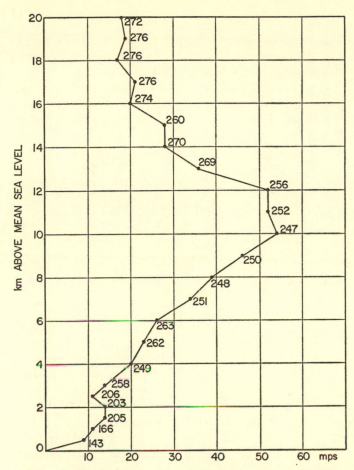

Fig. 15. Vertical wind profile measured over Green Bay, Wisconsin on 19 April 1963, 00 Greenwich Mean Time (or 18 April, 6 P.M. Central Standard Time). Numbers given along profile are wind directions in degrees.

On the North American continent there are approximately 100 such stations (Figure 16). All measurements

Fig. 16. Location of radiosonde stations of the United States and southern Canada.

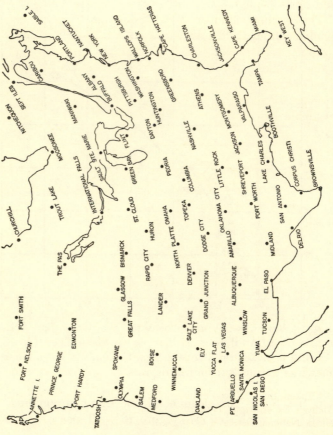

are made at the same time, at oo and 12 Greenwich
Mean Time; that is, when it is midnight and noon in
the location of the meridian that runs through the Greenwich Observatory (o degrees longitude). Eastern Standard Time, for instance, runs five hours behind Greenwich
Mean Time. Thus, the observations in this time zone are
made at 7 A.M. and 7 P.M. EST. Since the same observation times are kept over the entire globe, we call these
synoptic observations, from the Greek *syn* (together)
and *opsis* (sight).

The greatest number of upper-air observations are
available from the North American continent and from
Europe. A good many are taken also in Asia and Australia. In the North Atlantic and North Pacific there are
a few weather ships and island stations that make measurements regularly (Figure 17). Very little data are
available, however, from the rest of the globe.

Fig. 17. Location* of stations regularly reporting upper-air
data.

From a vertical wind profile alone, which gives wind
speed as a function of height, we do not see how the

* Taken from "The Feasibility of a Global Observation and
Analysis Experiment," Publication 1290, National Academy of Sciences, National Research Council, Washington, D.C. (1966).

wind turns with height and from which direction it blows
at each level. Sometimes it is of great interest for the
weather forecaster to know whether the wind *veers* or
backs with height; that is, whether it turns *clockwise* or
counterclockwise with increasing elevation. (For instance, if the wind turns from south to west, it *veers;*
if it turns from east to north, it *backs.*)

Such behavior of the winds may be studied from a
hodograph (Figure 18). This is a diagram in which con-

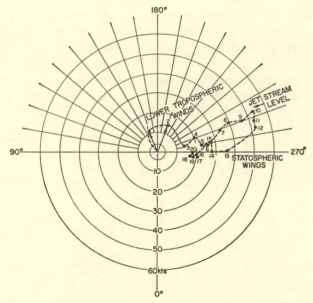

Fig. 18. Hodograph of wind sounding taken at Green Bay,
Wisconsin on 19 April 1963, 00 Greenwich Mean Time.

centric circles are drawn for wind speed (for example,
for every 10 *mps*), and radii are entered for every 10
degrees of wind direction. For each successive measurement level (1, 2, 3, 4, etc.) the wind vectors, extending from the center of the circles into the appropriate

wind direction, are entered. Note that the directions on
the hodograph are labeled the "wrong way," that is, 0°
or "north" is on the bottom, 180° or "south" is on the
top, etc. If we start to draw the wind vectors from the
center of the circle, a north wind (from 0°) will point
from north to south. Thus, the tip of a "north wind
vector" will lie along the line that is labeled 0°.

In Figure 18 the wind sounding of Green Bay, Wisconsin shown in Figure 15 has been reproduced. To keep
from cluttering up the picture, only a few wind vectors
have actually been drawn. The rest of them could easily
be constructed by connecting each point of the hodograph with the center of the diagram. The small numbers plotted along the hodograph give the height of the
wind reports in kilometers.

Comparing Figure 18 with Figure 15 we find that the
jet stream level may easily be identified from a hodograph. It is given by the level at which the wind vectors
"stick out farthest." We see furthermore that near the jet
stream level wind directions are fairly uniform. This uni-
formity seems to be a general characteristic of jet
streams. It is only in the lower troposphere and at some
distance above the jet-stream level that winds blow from
significantly different directions. In our example, winds
near the earth's surface blow from the southeast, veering into southwesterly winds at jet-stream level, then
veering further into westerly winds at stratospheric
heights.

Aircraft Measurements of Wind

If an aircraft is navigated properly, it can be used to
measure winds. The basic principles of navigation have
been described on pages 19–21. Suppose the pilot or
navigator observes carefully his true air speed (TAS)
and true heading (see Figure 4). Trying to keep both
constant, he obtains a fix on his geographic position once,

let us say, every hour. He can either compare landmarks with a map or measure the position of a fixed star or measure the angle of the aircraft's position with respect to two radio beacons (Figure 19). (The intersection of

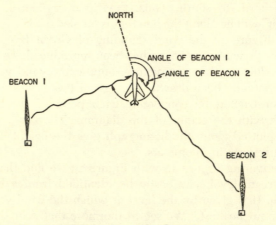

Fig. 19. Position of aircraft can be fixed by measuring angles at which radio signals are received from two beacons. Plane is at intersection of the two beacons.

the two beacons, plotted on a navigation chart, gives the position of the aircraft.)

The distance and direction between two successive fixes (F_1 and F_2) determine the *ground speed vector* because they measure the actual path which the airplane has traveled (Figure 20). If the two fixes have been obtained exactly one hour apart, then the ground speed may be read off directly in miles per hour or kilometers per hour from the scale of the map on which the navigator plotted the course of the aircraft.

Suppose that the *true heading* has been kept constant during this hour. It indicates the direction in which the aircraft *supposedly* was flying, had the navigator not

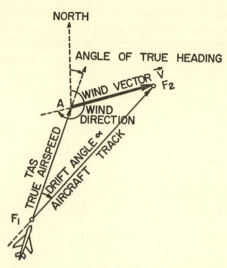

Fig. 20. Plane navigator can easily determine direction and magnitude of wind vector from fixes F_1 and F_2, one hour apart.

taken his geographic fixes. We enter the *true heading* as a line, starting in point F_1. Along this true heading the aircraft supposedly proceeded with the true air speed of so-and-so-many miles per hour, which the pilot tried to keep constant during this one hour of flight. If we enter the true air speed according to the scale of the map along the true heading line, starting at the point F_1, we obtain A. Theoretically, the aircraft should be over this point A after one hour of flight, starting to count time over point F_1. In reality, however, the airplane is over F_2. It drifted off because of the action of the wind.

The mean wind vector $\vec{V}$, which influenced the aircraft during this one hour of flight, is given by the length and the direction of the line from A to F_2. It may be read off immediately in miles per hour or kilometers per hour from the scale of the map.

From the various lines drawn on our navigation map, we can measure the drift angle α, caused by the wind, as well as the angle of *wind direction.*

With these explanations you should consider yourself a seasoned navigator. Would you trust yourself with the following example? An aircraft flies from Denver at a true heading of 80° and a true air speed of 200 knots. After exactly two hours, the pilot finds himself over Topeka, Kansas. Determine drift angle, wind speed in nautical miles per hour (knots), and wind direction in degrees.

From this example we may see that with this method of navigation the navigator will be the busiest man on board the aircraft if the plane is not to become lost. In cloudy weather it may be impossible to obtain a fix from landmarks or stars, and over the vast oceans, especially of the Southern Hemisphere, there are no radio beacons to guide the aircraft. The navigator will have to depend only on the forecast wind vector (which may not be very accurate) and on the indicators of heading and air speed on board the aircraft. We call this type of navigation *dead reckoning* because the navigator has no "live" fix available to check his computations. If the wind forecast is wrong, the aircraft may become hopelessly lost. This has happened.

Doppler Radar

A big breakthrough in navigation—and also in wind measurement by aircraft—came from the development of *Doppler radar.* The principle is based on the *Doppler effect,* which you have observed many times. If a train rapidly approaches an intersection, its whistle has a high-pitched sound. The moment the engine passes you the pitch of the whistle drops off sharply. Why?

The pitch of the whistle is determined by the frequency of sound waves, or by the number of waves per second, which hit your ear. The higher the pitch, the higher the frequency. If the train does not move, the sound waves will spread out as concentric circles, moving in every direction at the speed of sound, approximately 1000 feet per second (Figure 21a). If, however,

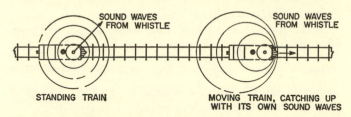

Fig. 21. Doppler effect, which raises pitch of approaching train whistle, results from "piling up" of sound waves within a given distance. To listener, result is the same as increasing frequency (or shortening wavelength) of the sound.

the train moves (Figure 21b), it is trying to catch up with the sound waves from its whistle in the direction of motion, and it is trying to run away from the waves that spread opposite to its direction of motion. Therefore, ahead of the train the successive sound waves are closer spaced than behind the train. Closer spacing means higher pitch.

Let us suppose that the whistle generates a sound of 1000 cycles per second. That means that 1000 waves per second will hit the ear. Since these waves will travel at 1000 feet per second, there will be 1000 waves filling the distance traveled in this one second. Thus, each wave will be one foot long. Suppose the train moves toward you at 50 *fps*. The 1000 sound waves generated during one second will now fill a space of only 950 feet. Thus, the wave length will be $\frac{950}{1000} = 0.95$ *ft*. The frequency is given as

$$\frac{\text{speed of sound}}{\text{wave length}} = \frac{(ft \text{ per second})}{(ft)}$$

$$= \frac{1000}{0.95} \text{, measured in units of } 1/sec$$

or 1053 cycles per second. Thus, the pitch of the whistle, compared with a standing train's whistle, is 53 cycles higher as the train approaches.

As the train speeds away, 1000 sound waves will now fill a space of 1050 feet, with a wave length of 1.05 feet and a frequency of $\frac{1000}{10.5} = 952$ cycles per second. The change in the pitch of the whistle as the train goes past you, therefore, is 101 cycles per second. If your ears were as sensitive as the ears of a trained musician, you could measure the speed of the train from the change in the sound of its whistle.

The Doppler radar does precisely what a trained musician could do in estimating the speed of a passing train. The only differences are that the plane is emitting electromagnetic pulses, which have a speed of 299,790 *km/sec,* and that the aircraft flies some distance above the ground. Therefore, it has to direct the beam downward at an angle which we may call γ (Figure 22).

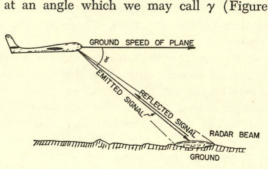

Fig. 22. Doppler radar measures ground speed of plane by comparing shift in frequency of the radar beam emitted from aircraft and the return beam received by reflection from earth.

With sensitive electronic equipment we measure the change in frequency between the emitted signal and the signal reflected from the ground. The signal will travel the distance between airplane and ground twice, so we have to count it twice in our computation.

The Doppler equation relates the frequency shift to the ground speed of the aircraft and to the *dip angle* γ of the radar beam:

Frequency Shift =
$$\frac{2 \times (\text{Ground Speed of Plane}) \times (\text{Frequency of Signal}) \times \cos\gamma}{\text{Speed of Light}}$$

In this equation ground speed and speed of light have to be expressed in the same units, for instance in *km/h*.

In astronomy the Doppler effect is used to estimate the speed with which distant stars travel away from the earth. We know that certain atoms will emit light of a certain color—that is, of a certain frequency—if these atoms are in a very hot environment. As a star moves away from us, its light will not appear at exactly the frequencies of atomic spectra measured in the laboratory, but a shift toward lower frequencies is observed. Since red light is characterized by relatively low frequencies, blue light by high frequencies, the light from a star approaching us will have characteristic bright bands in its light which are shifted toward the blue end of the spectrum. If the star travels away from us, the bands will be shifted toward the red end of the spectrum. Since we may assume that the light from the star travels along a straight line toward us, the angle γ will be zero, and therefore, $\cos\gamma = 1$. That means we don't have to worry about this factor any longer. By how much will a spectral line of light of 0.5 micron wave length be shifted, if the star emitting this line is traveling away at a speed of 10,000 *km/sec?* Clue:

$$\text{Frequency} = \frac{\text{Speed of light}}{\text{Wave length}}$$

$$1 \text{ micron} = 1 \ \mu = \frac{1}{1000} \ mm \text{ (millimeter)}$$

$$1 \ mm = \frac{1}{1000} \ m$$

With a sophisticated Doppler radar on board an aircraft, we not only measure the ground speed directly, but also the drift angle. This we do by sending out two radar beams, one to the left and forward, one to the right and forward of the aircraft. The reflected signal of each of the two beams will show a Doppler shift in its frequency. If the aircraft does not drift off to one side, the shift in both beams will be the same. If, however, there is a drift toward one side because of crosswinds, the beam toward which the aircraft is drifting will show a greater frequency shift than the other beam (Figure 23). The difference in frequency shift will be the larger, the larger the drift angle. Thus, we have a means of measuring this angle directly on board the aircraft.

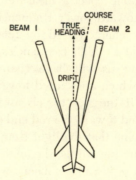

Fig. 23. Drift angle of plane in crosswind can be measured with two forward-pointing Doppler radar beams. Frequency shift will be greater in the reflected beam on the side toward which plane is drifting.

With the aid of Doppler radar one can compute the course of the aircraft by dead reckoning; that is, the navigator does not need to take a celestial or geographic fix of the location of the aircraft—provided, of course, the Doppler instrument is working properly.

Wind Shear

In the definition of jet streams on page 9 the term *wind shear* was used. Whenever the wind varies along a coordinate direction, shearing stresses will be caused; that is, the faster air will tend to drag the slower air along. We call this variation of wind over a given distance the *wind shear:*

Wind shear =
$$\frac{\text{Wind speed difference}}{\text{Distance over which this difference is measured}}$$

Suppose the wind at 100 m height is 20 mps. At the ground itself the wind speed is zero. What is the wind shear between these two levels? The answer is very simple. The shear is $\frac{20}{100} = 0.2$, in units of $1/sec$.

On page 36 we learned about vertical wind profiles. Whenever the wind is seen to vary in speed along such a profile, a vertical *wind shear* is present. It is only when the wind speed remains constant with height that there is no vertical wind shear because in such a case the wind speed difference between two levels will be zero. Suppose the wind speed measured 1000 feet above the ground is 30 mps, and at 2000 feet above the ground it is also 30 mps blowing from the same direction. Clearly 30 − 30 = 0 mps; thus, there is no wind shear between the two levels.

We have to be careful with this statement, however, because the wind is a *vector* defined by *speed and direction*. Suppose at 1000 feet we measure a west wind of 5 *mps;* at 2000 feet, we observe an east wind of 5 *mps*. Clearly, there is a wind shear present now, namely 10 *mps*.

As shown in Figure 24, things are more complicated

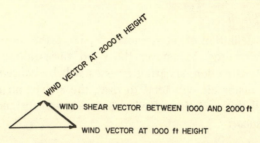

Fig. 24. Wind shear vector points from first wind vector of reference to second wind vector. Shear vector here shows that wind is backing with height.

if the wind turns with height by an angle different from 180°. The shear vector may be constructed as the line pointing from the 1000-foot wind vector to the 2000-foot wind vector. In the example of Figure 24 the wind is *backing* with height.

From Figure 18 we remember that wind vectors drawn as in Figure 24 are conveniently displayed in a *hodograph*. Thus, if we connect the end points of the vectors in a hodograph, we immediately arrive at the *shear vectors*. Their magnitudes may be measured in *mps* or knots, on the scale offered by the radial distance between the concentric circles of the hodograph. For example, between 12 and 13 *km* the shear vector is 19 *mps* per *km* (or 0.019 1/*sec*), and is directed at 51 degrees (from the northeast).

Horizontal shears are easily measured, too. If winds are plotted on a map as in Figure 25, all we have to do

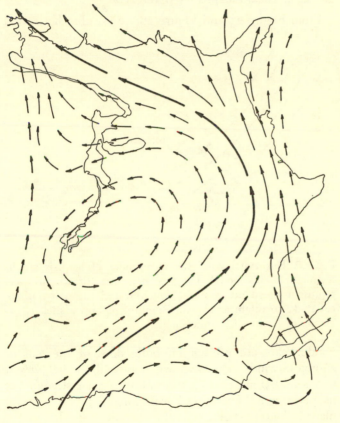

Fig. 25. Jet stream flow over North America is shown schematically on this map. Arrows, their lengths proportional to wind speeds, point in direction of wind. Wind speeds along jet axis (heaviest arrows) in winter exceed 120 knots.

is to determine the wind speed difference over some horizontal distance taken perpendicular to the direction of
the flow. We may use the map scale to measure off this
distance. Again

Wind shear =

$$\frac{\text{Wind speed difference}}{\text{Distance over which this difference is measured}} \text{,}$$

and again shear is expressed in terms of $\frac{1}{sec}$.

From Figure 25 and Figure 26 we see that horizontal

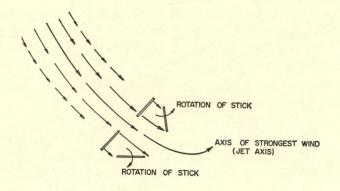

Fig. 26. Horizontal wind shears on sides of jet axis are il-
lustrated in this model. Lengths of arrows, drawn to scale,
represent speed differentials, which would cause hypothetical sticks to rotate as shown. Rotation of upper stick is cyclonic, of lower stick anticyclonic.

wind shears are present on both sides of the axis of
strongest winds. This axis we may call the *jet axis* if
winds are strong enough to qualify as a jet stream. Let
us perform a hypothetical experiment: If we were to
place a floating stick across the current to the left of the
jet axis (looking downstream), the horizontal wind shear

there would cause this stick to turn counterclockwise. The north end of the stick would swim slower in its surrounding slow winds than the south end of the stick. If we were to perform the same experiment on the right-hand side of the jet axis (looking downstream), the stick would turn clockwise.

Counterclockwise rotation in the Northern Hemisphere is called *cyclonic rotation;* clockwise rotation in the Northern Hemisphere is called *anticyclonic.* Therefore, we say that the wind shear left of the jet axis is *cyclonic;* right of the jet axis it is *anticyclonic.*

To help us grasp the phenomenon of a jet stream a little better the World Meteorological Organization has attached an additional explanation to its statement, which was quoted on page 9.

> Normally a jet stream is thousands of kilometers in length, hundreds of kilometers in width and some kilometers in depth. The vertical shear of wind is of the order of 5 to 10 *mps* per kilometer and the lateral shear is of the order of 5 *mps* per 100 *km.* An arbitrary lower limit of 30 *mps* is assigned to the speed of the wind along the axis of a jet stream.

Let us imagine a jet stream with 100 *mps* maximum wind speed and with its axis located 10 *km* above the ground. How deep and how wide would this jet stream be if we assumed the above values of mean vertical and horizontal wind shear, and if we assumed that only the areas with winds stronger than 30 *mps* belong to the jet stream?

With a vertical shear of 10 *mps* per *km,* we would calculate the jet stream in this example to be 14 *km* thick, assuming that the same linear shear holds in the troposphere as well as in the stratosphere. (The jet is at a height of 10 *km.* It reaches 7 *km* into the troposphere and 7 *km* into the stratosphere under the assumptions made above.) With a horizontal shear of 5 *mps* per 100 *km*, the jet stream would be 2 × 1400 *km* = 2800 *km* wide.

As we may see in Figure 15, vertical shears are not quite the same above and below the level of maximum wind. In the example of this figure, the jet stream is approximately 7 *km* thick. Usually cyclonic shears are stronger than anticyclonic shears, as we will see in later examples. This differential will somewhat reduce the width of jet streams below the above estimate. Nevertheless, we may see that typical jet streams are shallow "sheets" of high wind velocities embedded in the atmosphere.

Chapter IV

FINDING THE JET STREAM

How to "See the Forest for the Trees"

Though the jet stream may have been a mystery to the aviators of World War II, in this day and age of space travel we don't need a magnifying glass to search for clues concerning these "rivers of air." Jet streams nowadays are a component of the sterile atmosphere of a Weather Bureau Operations Office. Hundreds of measurements taken all over the globe each day keep continuous watch on their movements and behavior. Airline and military pilots supplement these measurements continuously with their own reports, thus adding valuable information over areas in which no radiosonde stations track the air motions.

In the preceding chapter we have described techniques of measuring upper winds and jet streams. Let us now apply the results of these measurements to a search for the jet stream. Actually, such a search is not as exciting (and not as frustrating) as a treasure hunt, because the clues, although in code, are easily deciphered into plain language. Let's start off with radiosonde and rawinsonde observations.

From the balloon positions obtained each minute of the sounding run the winds are calculated as described previously. Vertical wind profiles, depicting the distribution of wind speed and wind direction, are constructed. The information thus gained will now have to be compared with soundings made at other stations. This, as we may easily guess, poses a communications problem. If we make our wind measurements carefully enough, we will detect a large number of small fluctuations in the wind soundings. One instrument, for instance, which provides such great accuracy, is the FPS-16 tracking radar, used to follow the course of missiles through the sky. If we home this radar in on ascending balloons, we find an astonishingly turbulent state of the atmosphere. Figure 27 shows an example of such measurements, taken over Cape Kennedy every 45 minutes.

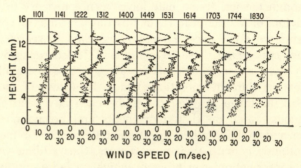

Fig. 27. Wind soundings taken with radar over Cape Kennedy on 3 January 1963, Greenwich Mean Time of balloon release is entered at top. Height in kilometers is plotted against wind speed in meters per second.

In looking at this diagram more closely, we find that some of the wiggles appearing in the vertical wind profiles persist for hours; others are more transient in nature and can be identified in only one or two ascents. The lower parts of the soundings appear to be fuzzy. The individual measurement points obtained every three seconds (instead of once every minute, as with regular sounding radar equipment) spread out into a thick line. Only at greater heights, over 10 *km* or so, do the individual dots fall into neat curves. The explanation is that the FPS-16 radar is almost too accurate for our meteorological purposes. In tracking a balloon it actually picks up erratic motions that result from the balloon's shedding little eddies from its skin as it rises through the air. (We can prove this by releasing a balloon in a perfectly wind-still hangar. It will not rise vertically, but follow a zigzag line upward, for reasons of self-generated turbulence. These erratic motions are picked up by FPS-16 radar.)

Clearly, in trying to identify jet streams we cannot pay any attention to balloon-induced turbulent motions. We cannot even pay attention to actual turbulence in the atmosphere, for its eddies are short-lived. They last, maybe, a few minutes at the most. Upper-air measurements are taken every 12 hours, however. Certainly the small-scale turbulence will not persist that long.

This brings us to a problem very essential to meteorology, but also to other fields of physical measurement. What accuracy and resolution are required from my measurements, in order to be able to measure what I am actually interested in?

Clearly, setting twelve-hour intervals between upper-air soundings is intended to detect atmospheric structure and motions on only a relatively large scale, such as influence the daily weather patterns, cyclones, anticyclones, fronts, and alike. If we made measurements every half hour, we possibly could follow and describe such persistent wiggles, as they appear in Figure 27. A finer resolution of time-scale by making more frequent observations would not solve every problem either. Radiosonde stations over North America, where we pride ourselves on having the densest network on the globe, are still about 300 miles apart. Even if we took soundings every few minutes, we would not know what was going on in between the stations. Thus, we would have to install a radiosonde station every few kilometers at least. But think how the taxpayers would react to this. Our meteorological observations under such a scheme would probably cost a considerable fraction of the gross national product.

Since such a proposal is obviously impractical, we have to learn to live with what we have, and with what we can easily obtain. Therefore, for purposes of large-scale weather analysis and forecasting, let's forget about the little details in the atmosphere. There is one drawback in doing this, however, and we had better be aware of the loss: In meteorology, more than in any other science, the saying holds, "Big trees grow from small seeds." Small, neglected details in atmospheric structure and flow patterns may, at some later time and farther downstream, trigger the development of big storms. Thus, being incapable of measuring everything everywhere all the time, we have to admit that our weather forecasts, no matter how good and sophisticated they are, will deteriorate in quality with increasing length of the time period over which they are issued. Even satellites and manned orbital platforms will not be able to alter this basic truth—no matter what a newspaper reporter tells you.

Since obviously we cannot be omniscient of what is going on in the atmosphere, there is nothing gained in retaining even the more interesting wiggles shown in our soundings of Figure 27. This neglect, fortunately, takes some of the pressure off our communications problem. For now we will be justified in making a limited selection of points along the wind soundings—the so-called standard levels—for which wind values along the vertical wind profiles can be put into digital form and transmitted to a central office for further processing.

Those who have retained a certain amount of buccaneering spirit may find interest in the fact that these transmissions follow a certain "secret" code, which allows for date and time, station number, etc. Different codes are used for wind reports only, and for winds measured together with temperature, pressure, and humidity by a regular radiosonde. A full explanation is given in Table

II. As an example, Table III contains actual sounding reports from Green Bay, Wisconsin (Station No. 645), for April 19, 1963, oo Greenwich Mean Time. This report is plotted in a vertical wind profile in Figure 15. The temperature sounding may be reproduced in a diagram which contains temperature and pressure—decreasing in a non-linear (exponential) way—as coordinates (Figure 28). (Pressure decreases with height, as the air becomes

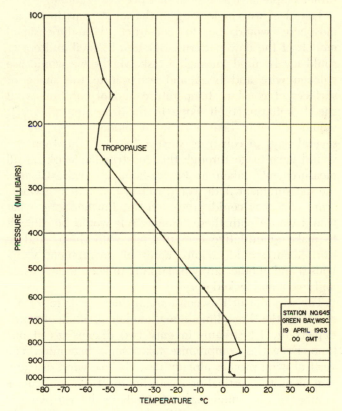

Fig. 28. Temperature sounding over Green Bay, Wisconsin, 19 April, 1963, oo Greenwich Mean Time.

thinner and thinner. The rate of decrease, however, slows down as we go higher up. These facts are incorporated in an exponential function.)

Mapping the Jet Stream

The coded information on winds from all sounding stations will now have to be compared with each other. This may sound like a formidable task, but it is actually rather simple once a "system" has been worked out.

The ideal way of looking at the jet streams in the atmosphere would be to construct three-dimensional models of the flow patterns. But how? Pencil and eraser could not be used for such a task. Maybe we would use chicken wire and bend and warp it to the shape of surfaces of constant temperature and of constant wind speed, imitating their slope in the free atmosphere. It would probably take a team of modern "pop" artists several days to construct such a "model." But then, we would have to go through this construction work for all measurements taken at twelve-hour intervals. Clearly, the measurement data would roll in faster than a whole army of artists could process them. Furthermore, what would we do with these "models"? It would be difficult to make quantitative computations with them. And to store them would require acres of government warehouses and another army of janitors to keep them dusted and swept and locked up.

The way out of this nonsensical dilemma is to let our minds do the three-dimensional modeling, while our "art" depicts the jet streams in only two dimensions at one time. Even this compromise takes some ingenuity. The vertical wind profiles already describe the wind distribution along the vertical or z-coordinate. If we construct a number of horizontal maps of winds in the x-y plane at different heights z, we will have a poor but usable substitute for a three-dimensional presentation (Figure 29).

Codes for Upper Air Observations from Land Stations

Wind Reports:

IIiii GGi_hDf_a Hddff   Hddff . . .   9999n   Hddff . . .

Radiosonde Reports:

IIiii $GGh_1h_1h_1$ $T_1T_1T_{d1}T_{d1}T_{x1}$ $od_1d_1f_1f_1$

$P_2P_2h_2h_2h_2$ $T_2T_2T_{d2}T_{d2}T_{x2}$ $od_2d_2f_2f_2$

- - - - - - - -   - - - - - - - - - -   - - - - - - - -

55555 $ooP_oP_oP_o$ $T_oT_oT_{do}T_{do}T_{xo}$ $od_od_of_of_o$

$n_1n_1P_1P_1P_1$ $T_1T_1T_{d1}T_{d1}T_{x1}$ $od_1d_1f_1f_1$

Explanation:

Wind Reports:

II Block number, indicating area from which the station is reporting. Example: Alaskan stations report under Block 70, continental United States and Canada under Blocks 72 and 74, Mexico under Block 76, England and Ireland under Block 03, France under Block 07, Germany under Block 10, the U.S.S.R. under Blocks 20 through 38.

iii Station number. Example: Green Bay, Wisconsin has station number 645.

GG Observation time, given in Greenwich Mean Time (usually 00 or 12 GMT).

i_h Code number, indicating the height intervals at which the following wind reports are given, and the type of instrumentation used in measuring winds (pilot balloon or radio theodolite).

D Direction of surface winds, measured in an 8-point wind rose. Example: "0" means calm; "1" means northeast winds; "2" means east winds; "4" means south winds; "8" means north winds; "9" means variable winds.

f_a Surface wind speed in tens of knots or in units of 5 *mps*. Example: "0" means 0–4 knots or 0–2 *mps;* "1" means 5–14 knots or $2\frac{1}{2}$–7 *mps,* etc.

H   Height in units of 300 *m* (1000 *ft*), 500 *m,* or 1000 *m* above ground, depending on number given for i_h.

dd   Wind direction in tens of degrees of a 360° wind rose. Example: "24" means a wind from 236° to 245° (i.e., from southwest).

ff   Wind speed in knots. If wind is stronger than 99 knots, a value of 50 is added to the wind direction. Example: 88212 in this group would mean: at level 8 a wind direction of 320° and a speed of 112 knots has been measured.

$9999n$ n gives the tens that have to be added to H in the following groups.

Radiosonde Reports:

Symbols not explained here are the same as for wind reports.

$h_1h_1h_1$ Height of standard pressure levels 1, 2, etc. above $h_2h_2h_2$ sea level, given in meters (leaving out thousands etc.   and ten thousands) or in tens of feet (leaving out units, ten thousands, and hundred thousands).

T_1T_1 Temperatures at pressure levels 1, 2, etc., in de- T_2T_2 grees Centigrade. For negative values between 0 etc.   and −50° a value of 50 is added. Example: "12" could mean +12° C or −62° C; "78" would mean −28° C.

$T_{d1}T_{d1}$ Dew point temperatures at pressure levels 1, 2, $T_{d2}T_{d2}$ etc., defined as the temperature at which the

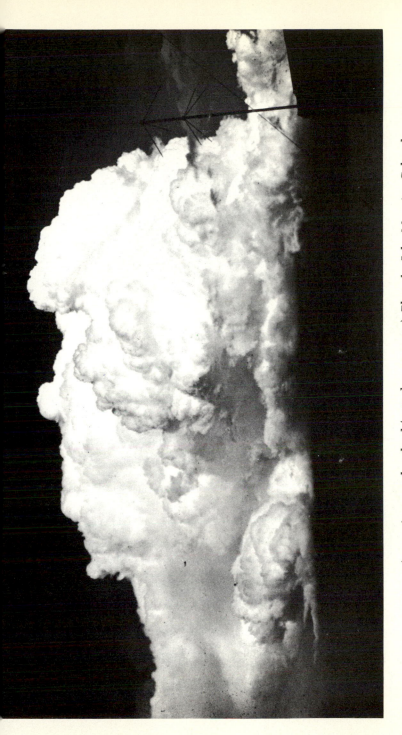

PLATE I. A towering cumulus cloud (*cumulus congestus*). Photo by John Marwitz, Colorado State University.

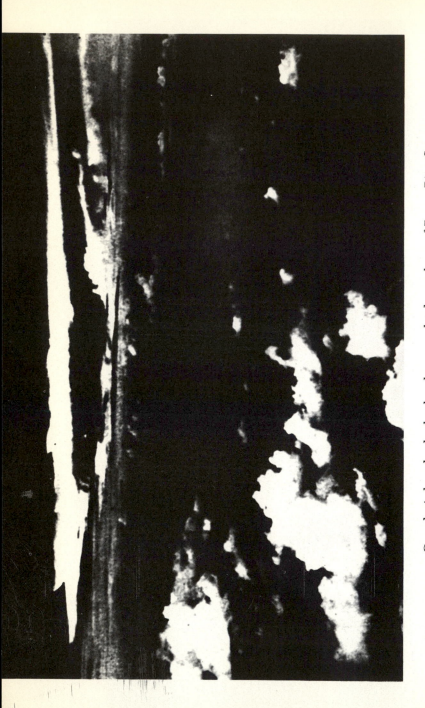

PLATE II. Cumulonimbus clouds, the thunderstorm clouds, northeast of Puerto Rico. Strong vertical wind shears near tropopause are blowing off the anvils. Photo by J. S. Malkus.

PLATE III. The device that looks like a lampshade is a radiation shield in which sensitive thermometers are installed. Photo by National Bureau of Standards, Boulder, Colorado.

PLATE IV. Radiosonde readied for launch is held by man at right. The dish on the tripod is antenna of radiosonde receiver. In foreground, balloon is connected to helium bottle for inflation. Photo by Duayne Barnhart, Colorado State University.

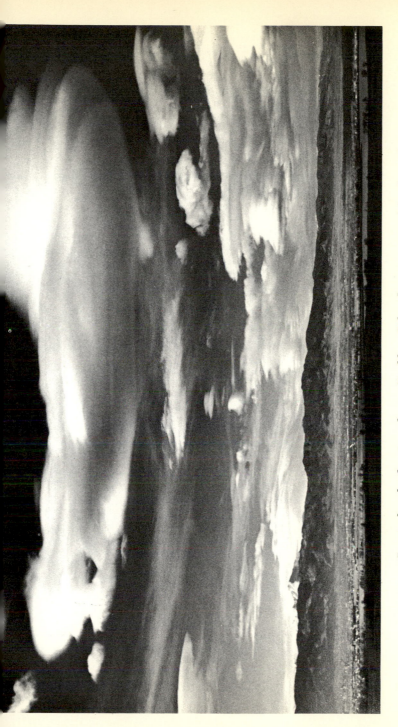

PLATE v. Lenticular clouds, viewed near Boulder, Colorado, were caused by flow of air over Continental Divide. Photo by National Center for Atmospheric Research, Boulder.

PLATE VI. Tetrahedron reflector of aluminum foil will be carried aloft by sounding balloon. Photo by Duayne Barnhart, Colorado State University.

etc. moisture of the air—if cooled down to this value —would reach saturation.

T_{x1} T_{x2} etc. Code number, allowing temperature and dew point temperature to be estimated to the nearest 0.3° C.

o Code number, indicating that the following four digits give wind direction in tens of degrees and wind speeds in knots or *mps* (see explanation of wind reports).

P_2P_2 P_3P_3 etc. Standard pressure levels in tens of millibars. Example: "85" means 850 *mb*, "25" means 250 *mb*.

55555 Group of code numbers, signaling the reports of significant points along the sounding are following at which the vertical temperature or moisture profile shows marked "kinks."

oo n_1n_1 etc. The first, second, etc., such significant point follows.

TABLE III

Example of Coded Message for Green Bay,
Wisconsin, April 19, 1963, oo Greenwich Mean Time
(See Figures 15 and 28)

Wind Report:

72645	00051	11722	22028	32628	42540
	52646	62652	72568	82578	92592
	99991	07506	17504	27504	32772
	42756	52656	62740	72842	82834
	92838	99992	02736		

Note that in Figure 15 speeds have been plotted in *mps*, whereas the coded message above contains wind speeds in knots.

Radiosonde Report:

```
72645    00078    85452    09541    02013
                  70975    50535    02513
                  50590    66690    02630
                  40231    77819    02539
                  30229    93995    02545
                  25430    03990    02554

         55555    00984    05520    01204
                  11969    02639    01409
                  22873    03591    02014
                  33851    07019    02013
                  44575    58595    02622
```

Note   The 1000 *mb* level, reported in the second group of 5 digits and constituting the first standard pressure level, lies below the terrain (78 *m* above means sea level). Therefore, no temperatures and winds are reported for this level. The third group of 5 digits, thus, gives the 85 *mb* height of 1452 *m*. (Heights in this coded message are given in meters, winds in meters per second, although in the United States presently feet and knots are still the customary units.)

Other heights are   700 *mb* – 2975 *m*
                      500 *mb* – 5590 *m*
                      400 *mb* – 7231 *m*
                      300 *mb* – 9229 *m*
                      250 *mb* – 10430 *m*

The first significant point gives the surface pressure of 984 *mb* at the station, the associated temperature and dew point temperature, and the surface wind report.

Instead of drawing maps on surfaces of constant height, meteorologists have become accustomed to analyzing winds and temperatures on surfaces of *constant pressure*. They talk about 850 *mb* maps, 500 *mb* maps, and 300 *mb* maps, etc., meaning that whatever parameters are shown on these maps occur at a level in the atmosphere where the pressure measures 850, 500, or 300

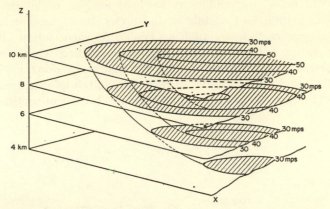

Fig. 29. Usable (if poor) substitute for a true three-dimensional representation of a wind field is this stack of two-dimensional plots of wind speeds at regular intervals of altitude.

millibars, respectively. Airlines have found a certain advantage in this, because airplanes don't usually fly at constant geometric altitude (above mean sea level), but at constant *pressure altitude*, as indicated by the *altimeter*. This device is nothing but a barometer, calibrated in feet above sea level rather than in inches of mercury, indicating the height of a certain constant-pressure surface. If this surface is "warped" in space, the airplane will climb up and down, following this surface and maintaining its constant "pressure altitude" at the same time.

Now that we have decided to use constant-pressure maps to analyze the jet stream, our task is made easy. We take wind, temperature, height, and (if available) humidity data from the soundings at each station for one particular observation time, say April 19, 1963, 00 hours Greenwich Mean Time, and for one particular pressure level, say the 250 *mb* surface. These data are plotted on a map in which the locations of radiosonde stations are shown (see Figure 16).

In order to facilitate subsequent analysis of all this information on the state of the upper atmosphere, it is convenient to use a *plotting model*. At each station all data are entered in the same fashion so that when we proceed to analyze winds or temperatures, we don't need the mind of a detective to read the plotted information. Figure 30 shows such a plotting model, used in routine

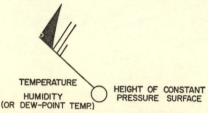

TEMPERATURE
HUMIDITY   HEIGHT OF CONSTANT
(OR DEW-POINT TEMP.)   PRESSURE SURFACE

Fig. 30. Plotting model displays standardized symbols for transcription of data on weather maps. Circle gives station location and is tip of "arrow" showing wind direction. "Barbs" of arrow tail show wind speed: fifty knots for each black triangle, ten for each longer dash, five for each short dash. Wind given here is from the northwest at seventy-five knots.

weather analysis. As a practical example, the 250 *mb* information for the radiosonde station of Green Bay, Wisconsin, April 19, 1963, 00 GMT, is given in Figure 31. The wind direction is indicated by the direction in which the arrow flies, the circle marking the location of the radiosonde station on the map. The wind speed is symbolized by the barbs in the tail of the arrow: One black triangle for each 50 knots, one long dash for each ten knots, and one short dash for each five knots. The example in Figure 30, thus, reads 75 knots; in Figure 31 it reads 55 knots.

Information on temperature and height of the respective pressure surface (in our case the 250 *mb* surface) is entered alongside the station numerically. Since the number for heights, expressed in feet or meters, may be

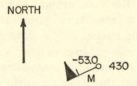

Fig. 31. Symbols here on a 250-*mb* map show a fifty-five-knot SW wind over Green Bay on 19 April 1963.

rather long, thus using up a lot of space on our relatively small map, one usually leaves out the first and last digits. The latter may be omitted, because the height measurements are only accurate to the nearest 10 feet or 10 meters anyway; therefore, there would be no point in reporting the unit digits. The ten-thousand digit may be omitted from plotting because the height of a constant pressure surface does not vary so much from equator to pole that a trained meteorologist would not know the number that has been dropped. In the case of the 250 *mb* surface, for instance, we know that it is located roughly near 10.5 *km*. Thus, if we omit the first number 1, we save some space in plotting, and nothing has been sacrificed in accuracy.

All these data, say for April 17, 1963, oo GMT, may now be assembled on a map of North America. In Figure 32 only wind arrows have been plotted. Temperature and height information has been left out in order to avoid crowding of the small map. What next? Looking at this diagram we find that the little wind arrows have a tendency to streak into certain directions of flow. They are not scattering wildly all over the place, but are outlining very definitely certain consistent rivers of air, with eddies branching off on the sides. We might think of these arrows as giving an instantaneous "flashlight" photograph of air motion at the 250 *mb* surface.

We may now perform the first and easiest step of analysis: Imagining that the plotted wind directions are

representative of the area around each radiosonde station, we can draw "streamers" parallel to the wind, indicating the configuration of the flow pattern. This has
been done in Figure 32. In order to sound more professional we will call these lines *streamlines*. We should be
careful to note that the streamlines do *not* connect successive radiosonde stations, but are everywhere *tangent*
to the wind direction in the particular places.

From the streamline pattern in Figure 32 we detect a
cyclonic vortex over the northwestern United States. A
cyclonic vortex (page 53) is one which rotates counterclockwise in the Northern Hemisphere, you will recall.
An *anticyclonic* eddy is located over Alberta and Sas-
katchewan. Since cyclonic motion is found around low-
pressure centers, anticyclonic flow around high-pressure
regions, we may immediately draw conclusions about
the pressure field from our streamline analysis: A low is
located over the Pacific Northwest, a high over Canada.

Next we may analyze the height of the 250 *mb* surface
from the values available from each radiosonde station.
This analysis is shown in Figure 33. In the region of
southwesterly winds, blowing across the United States,
the 250 *mb* contour lines run closely parallel to the
streamlines in Figure 32, indicating that the winds are
nearly *geostrophic**: The deflecting force of the earth's
rotation (which will be discussed in more detail in Chapter V) prevents the air from flowing directly from high
to low, but forces it to deviate toward the right in the
Northern Hemisphere. Your curiosity as to why this is so
will be satisfied later in this book. For now, let us just
accept the facts, by comparing Figures 32 and 33 with
each other.

A word is still in order on how we obtained this contour analysis. Very simple: The height values of the 250

* From the Greek: *geo,* earth; *strophein,* to deflect.

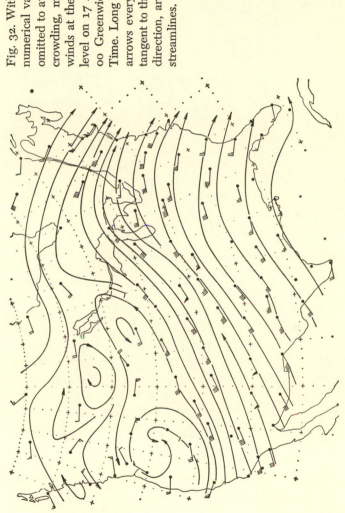

Fig. 32. With numerical values omitted to avoid crowding, map shows winds at the 250 *mb* level on 17 April 1963, 00 Greenwich Mean Time. Long lines with arrows everywhere tangent to the wind direction, are streamlines.

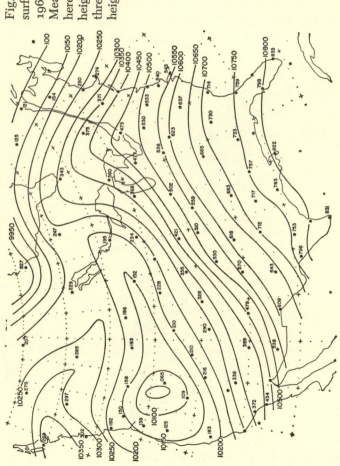

Fig. 33. The 250 *mb* surface on 17 April 1963, 00 Greenwich Mean Time, is shown here by contour heights in meters. Last three digits only of heights are entered.

mb isobaric surface,† which are plotted at each station (Figure 33), are interpolated for even 100-meter increments. Smooth lines are drawn for these 100-meter contour intervals. What we end up with is a geographic map, showing the "relief" of the 250 *mb* surface, very much as a contour map of the Rockies would show the relief of the mountains. In regions of anticyclonic flow the contours show high values; in regions of cyclonic flow they show low values. If, instead of mapping an isobaric surface by contour lines, we constructed isobars on a constant-height surface (let us say the 10.5 *km* level) in the cyclonic vortex over the northwestern United States, the 250 *mb* surface would fall below this level; thus pressure would be lower than 250 *mb* at the 10.5 *km* level. In the anticyclonic vortex over the Gulf of Mexico, for instance, the 250 *mb* level lies above 10.5 *km;* therefore, the 10.5 *km* pressure would be more than 250 *mb* there. Thus, we are allowed to call the "depression" or "valley" on our 250 *mb* contour map a *low* and the "mountain" or "ridge" a *high,* just as though we had analyzed pressures and not contours.

Now comes the exciting moment of our jet stream hunt. From the little barbs on our wind arrows in Figure 32 we have an indication of wind speed. We may now draw lines of equal wind speed—so-called *isotachs*‡—by interpolating between speed values reported at individual stations. This has been done in Figure 34 where we see the first "footprints" of a jet stream in the shape of elongated streaks of high wind velocity blowing across the country. (According to the definition given by the World Meteorological Organization, anything stronger than 60 knots or 30 *mps* could be considered a "jet stream." Areas meeting these qualifications have been shaded.)

† From the Greek: *isos,* equal; *baros,* weight.
‡ From the Greek: *isos,* equal; *tachos,* speed.

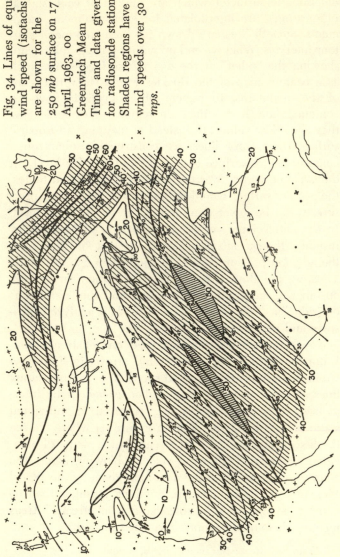

Fig. 34. Lines of equal wind speed (isotachs) are shown for the 250 *mb* surface on 17 April 1963, oo Greenwich Mean Time, and data given for radiosonde stations. Shaded regions have wind speeds over 30 *mps.*

In comparing the analytic steps we have taken so far, we may reach several very important conclusions which touch at the heart of jet-stream meteorology. First of all, we see that the jet stream is not one uniform band of high speeds, running from west to east in an orderly fashion. Note, for instance, the branch of high wind speeds that comes from northern Canada and joins the main flow off the New England coast. Even the main flow itself is split into several branches. Down the drain goes the idea of a simple, straightforward "model" of the atmospheric circulation, conceived by early researchers who started to follow up on the weird reports of pilots in World War II. Here again, as so many times in physics and meteorology, we discover that there are more things between heaven and earth than a simple textbook mind can fathom. Nature, it seems, has a split personality. On the one hand it stands out in brilliant simplicity, humbling itself into a few short mathematical equations; on the other hand, it baffles our imagination with a truly infinite variety of details, far beyond the grasp of even the most sophisticated electronic computers. Between these two poles of simplicity and diversity the searching mind of man remains caught: No matter what he does, or what he invents, there is always nature, ready to teach him a lesson in humility.

A few more facts may be seen from our analyses: Wind speeds are high where contour lines are close together, indicating a steep slope from the high-pressure "mountain" to the low-pressure "valley." This characteristic, again, conforms to the action of the deflecting force of the earth's rotation, which we will talk about later. Furthermore, the air motions are not all parallel, or side-by-side. The streamlines show regions in which the flow converges, others where it diverges. Together with the accelerations and decelerations which the air must undergo as it passes through the jet maxima (that is, centers of high wind speed along the jet streams), such convergence and divergence will cause air from the 250 *mb* level to be pushed out of the way into other pressure levels, or to be pushed into this level from adjacent layers above or below. In other words, our air motions will not be confined to the 250 *mb* level alone. There will be interactions with other pressure levels and the atmospheric flow on these levels, by means of vertical motions. These influences will cause the weather associated with jet streams. Again, you will have to hold your curiosity as to the how and why for a later chapter. We are not quite ready yet to discuss these implications.

There is still one analytic task remaining. We have not yet looked at the temperature reports available from each radiosonde station. Figure 35 shows a map of isotherms,§ arrived at by interpolation between the temperature values reported by individual soundings at the 250 *mb* pressure level. Isotherms are drawn for every two degrees Centigrade. The resulting pattern is strange, to say the least. We do not find a simple south-to-north temperature decrease, except over the southern United States. Near the jet stream streaks, and to the north of them, temperatures actually increase with latitude. In the discussion given in Chapter VI such a temperature distribution is held to be characteristic of the stratosphere. We may conclude, therefore, that the 250 *mb* level is located in the troposphere over the southern United States. It crosses into the stratosphere near the jet stream, and remains at stratospheric levels north of the jet stream.

Temperatures are especially high north of the jet maximum. As we will see later, the tropopause in these regions is located at particularly low levels. Some people have even talked about a "tropopause funnel" associated

§ From the Greek: *isos*, equal; *therme*, heat.

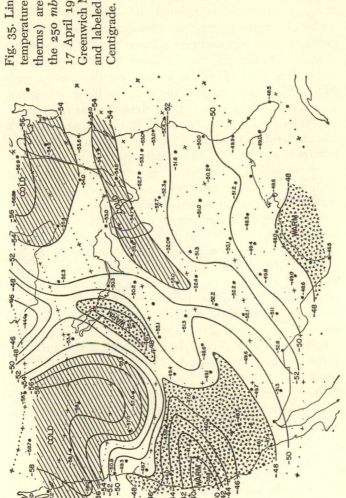

Fig. 35. Lines of equal temperature (iso-therms) are shown for the 250 *mb* surface on 17 April 1963, 00 Greenwich Mean Time, and labeled in degrees Centigrade.

with jet maxima. This notion has to do with sinking movements in the atmosphere, caused by the jet stream, and
of great importance to weather development. More details will be discussed in Chapter VII.

As we have now completed our analysis for a single
set of observations, we may henceforth make it a little
easier on ourselves and take somebody else's analyses for
granted. Figures 36 and 37 show isotach patterns for
April 18 and 19, 00 GMT. Comparing these maps with
Figure 34, we may deduce the behavior of the jet stream
over a forty-eight-hour period. Here are the facts:

The jet stream does not stand still in space; instead it
moves along. The whole jet stream pattern shifts gradually from west to east; so do individual jet maxima. They
show a general tendency to move along the *jet axes*,
indicated by heavy lines with arrows at their ends, de-
noting the axes of strongest winds along the direction of
flow. The jet axes themselves, however, also show some
movement. Let us pick an example. On April 17 (Figure
34) a small 40 *mps* wind maximum is located over Nevada. On the eighteenth the winds in this maximum have
increased above 50 *mps*, and the high-speed center is
now over Colorado. On the nineteenth this jet maximum
has merged with the main flow over the Great Lakes region. A displacement of approximately 10 degrees of
longitude per day toward the east would characterize the
speed of this small system. The jet maximum that appears over the Mexican border on the eighteenth, "wet-
backing" it into the United States, also has a similar rate
of displacement. It may be found over Oklahoma on the
nineteenth.

Ten degrees of longitude per day in middle latitudes
is close to a speed of 20 knots or 10 *mps*. If a jet maximum travels at an average speed of 10 *mps*, the wind
speed in the maximum may, however, be as high as 60
or 70 *mps*, and it clearly follows that the air has to ac-

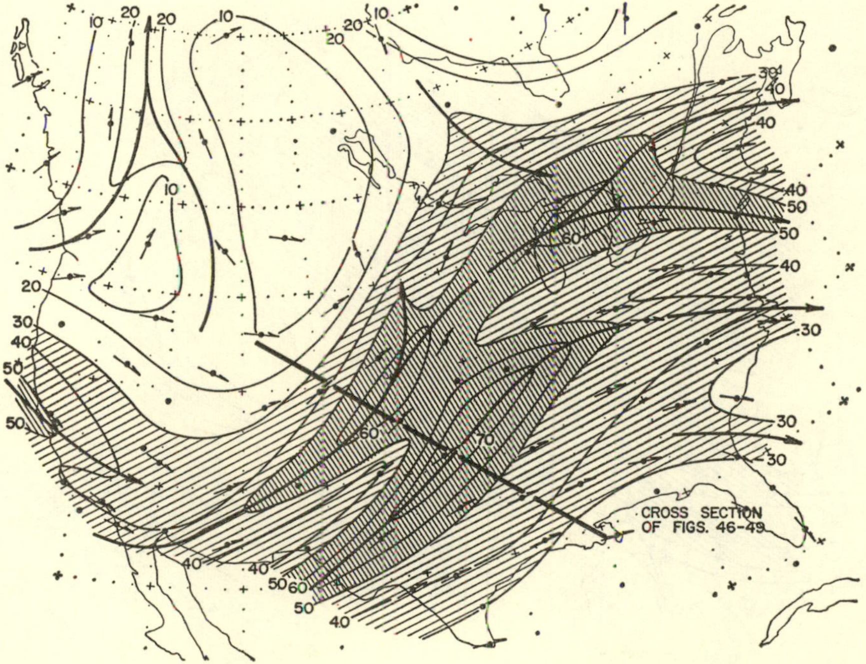

Fig. 36. Isotachs and wind arrows for the 250 *mb* surface of 18 April 1963.

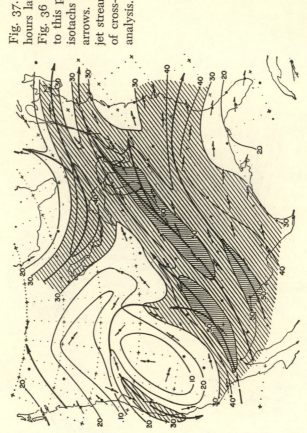

Fig. 37. Twenty-four hours later situation of Fig. 36 had changed to this picture of isotachs and wind arrows. Solid line cuts jet stream for location of cross-section analysis.

celerate as it enters the jet maximum from the rear, and decelerate as it shoots out in front. A jet maximum, thus, might be compared with a king-sized bagpipe. Air accelerates into the pipe as the piper plays his tune; at the same time, the piper marches on.

We also can identify several large meanders in the jet stream flow of Figures 34, 36, and 37. A ridge of high pressure is centered over the midwestern United States, while a trough of low pressure is situated off the United States east coast near Newfoundland. Going around the hemisphere, we find a whole succession of such troughs and ridges (Figure 38). Some of them are relatively short and move eastward with a speed of approximately 20 knots. Some of them are rather flat and wide and seem hardly to be moving at all. The latter we call quasi-stationary long-wave troughs and ridges. They seem to be linked to the distribution of large mountain ranges, such as the Himalayas and Rocky Mountains. The migrating short-wave troughs and ridges, on the other hand, are tied to such jet maxima as we have seen in Figures 34, 36, and 37. They cause the changes in weather which we observe from day to day. Before we get too deeply involved in a discussion of these problems, let us consider some other aspects.

We have stated before that a streamline analysis represents something like a flash photograph of the instantaneous flow configuration. Although we have not done so, we might draw streamlines for the maps of April 18 and 19 as well, and arrive at more instant photos, each slightly different from the previous one. If they were all the same, meaning that the flow pattern at jet-stream level did not change with time, the streamlines would indicate the actual displacement of air. This, however, is clearly not the case. The flow patterns change from day to day. While the air tries to move as the streamlines indicate it for the seventeenth of April, the same air gets

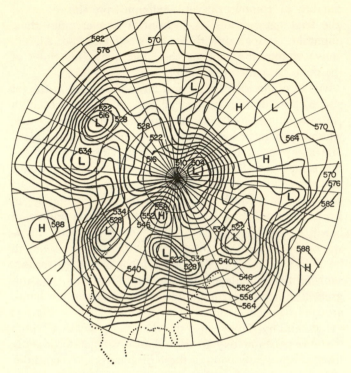

Fig. 38. Close spacing of contour lines on the 500 *mb* surface of 19 April 1963 indicates high wind velocities. Contour heights are given in tens of meters. Existence of jet-stream belt at middle latitudes, meandering across Asia, the Pacific, the United States (shown in dotted outline) and the Atlantic can be deduced.

caught in a slightly different flow configuration on the eighteenth, and so on. Suppose we could paint a small colored dot on a parcel of air and follow it on its journey, day after day. We would discover that the air actually does not follow the streamline pattern. The discrepancy between streamlines and the actual course of the air parcel becomes greater as the wind speed becomes smaller in comparison with the displacement speed of the flow pattern, and the difference between the wind direction and the direction in which the jet maxima move increases. We call these lines along which an air parcel actually travels the *air trajectories*. An example is given in Figure 39.

Computing air trajectories is quite a tedious task. In approximation, we may start out by assuming that an air parcel, located on the seventeenth of April, oo GMT on the 250 *mb* surface, will follow this streamline pattern for the next six hours. Then it will shift over to follow the streamlines of the seventeenth of April, 12 GMT, for the next twelve hours, the streamlines of the eighteenth, oo GMT, for the subsequent twelve hours, and so on. Clearly, this method of constructing trajectories is quite crude, because the flow patterns do not change suddenly at midnight or noon, but gradually throughout the day. Furthermore, the air parcel may not stay on the 250 *mb* surface all the time, but may be subject to vertical motions carrying it away from this surface into a different wind regime. Refinements of trajectory calculations, taking into account these possibilities, may become involved. But still, with the crude method outlined, we are able to arrive at a gross estimate of large-scale air motions.

To know such motions may be of considerable importance at times. Let us suppose that some accident wrecks a nuclear power plant, scattering poisonous fission products into the air. Meteorologists will now have to find out where this contaminated air will go. Will it hit populated regions? Will it drift into a neighboring country? To answer these questions calls for accurate trajectory analyses and forecasts.

Slicing through the Jet

So far we have developed a picture of the jet stream on *one* quasi-horizontal plane, stretched between the *x-y*

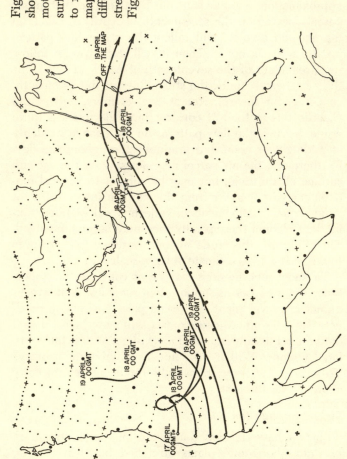

Fig. 39. Trajectories showing actual air motion at the 250 *mb* surface from 17 April to 19 April 1963 are mapped here. Note difference from streamline pattern in Fig. 32.

coordinate directions. How about the jet stream structure in the vertical? To gain an adequate view, meteorologists use so-called *cross sections* as an analytic tool. They are easy to construct.

Before we start out we should consider the following: In comparison with horizontal distances characteristic of jet streams (Figure 34) the atmosphere is only a thin "skin" surrounding the earth. A cross section through the jet stream would have to extend about 1000 miles in a direction approximately normal (that is, perpendicular) to the flow if it were to include all the characteristics of horizontal wind shear. The tropopause, however, is to be found at a height of maybe six or seven miles, less than 1 per cent of the distance covered by our cross section. If we used a sheet of graph paper, ten inches wide, let us say, to draw our cross section on, the jet stream, located near the tropopause, would lie $\frac{1}{10}$ of an inch above the bottom line representing the earth's surface. If we wished to analyze the flow and temperature structure in the troposphere on such a diagram, we would need an extremely sharp pencil and a magnifying glass, and even then we might not be successful. If a cross-sectional analysis is to be of any use at all, the vertical dimensions will have to be exaggerated tremendously, maybe fifty times or more, in comparison with the horizontal dimensions.

In constructing a cross section we start by drawing a straight line approximately normal to the flow, running close to as many radiosonde stations as possible (see Figure 37). The relative distance of these radiosonde stations along this line is then entered (with an appropriate scale factor) along the bottom line of the cross-section diagram (Figure 40).¶ The vertical coordinate

¶ Here we have chosen a chain of stations from Lander, Wyoming to Peoria, Illinois.

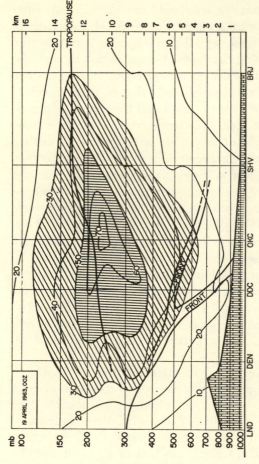

Fig. 40. Cross section along the heavy line of Fig. 37 is shown here. Isotachs in meters per second are analyzed with shadings to mark regions of speeds over 30 *mps* and 50 *mps*. Shading along bottom of map indicates terrain elevations for the stations from Lander, Wyoming (LND) through Denver (DEN), Dodge City (DDC), Oklahoma City (OKC), Shreveport (SHV) to Burrwood, Louisiana (BRJ). Heavy lines bound frontal zones and tropopause.

in this diagram shows pressure decreasing exponentially, as it does under average conditions in the atmosphere. Height in *km* or feet may also be given along the ordinate. We have to be aware, however, that the height scale holds only for *mean* (or standard) conditions. If the *real* atmosphere is too cold or too warm, the *actual* heights will show a discrepancy against the values printed alongside the rim of the cross section. To the aviator these discrepancies are known as the *altimeter corrections*. They tell him by how much the reading of his altimeter deviates from the actual height at which his aircraft is flying. Since meteorologists are used to constructing their cross sections with respect to pressure as the vertical coordinate, and not with respect to height above sea level, we will disregard these altimeter corrections.

After having located the position of the sounding stations along the bottom of the cross-section diagram, we may enter, along a vertical line above each station, the numbers for wind speeds, wind directions, temperatures, and humidities at the appropriate pressure levels at which they have been measured. All we have to do now is to analyze these values by interpolating between stations. In Figure 40 isotachs (lines of equal wind speed) have been drawn for each 5 *mps*. Figure 41 shows isogons (lines of equal wind direction)** for every 10 degrees of wind direction. Isotherms (lines of equal temperature) are drawn for every five degrees Centigrade in Figure 42.

Let us now take a comparative look at these diagrams. The *core* of the jet stream can be identified easily. Wind speeds in excess of 70 *mps* are found near 250 *mb* over Oklahoma City, in good agreement with the horizontal map shown in Figure 37. The three jet stream streaks or

** From the Greek: *isos*, equal; *gonia*, angle.

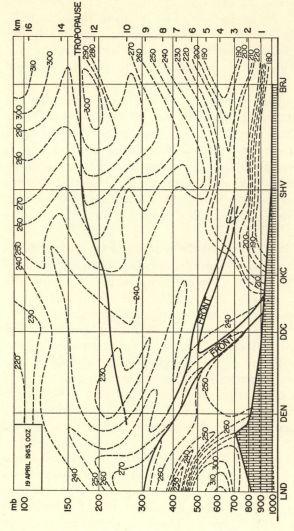

Fig. 41. Lines of equal wind direction (isogons) are analyzed for the cross section of Fig. 40, with the labels giving directions in degrees.

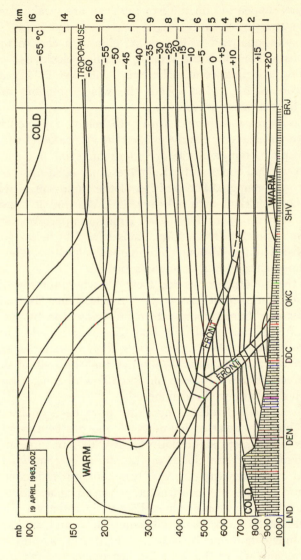

Fig. 42. Isotherms, in degrees Centigrade, are shown for the cross section of Fig. 40.

"fingers" appearing on the 250 *mb* map are weakly evident also in the cross section.

Isogons, analyzed in Figure 41, give us a clue to the regions where winds are veering (turning clockwise) or backing (turning counterclockwise with height). Near the core of high-wind speeds the wind directions are not changing very drastically.

The decrease of temperature with height, expected in the troposphere, is clearly evident in Figure 42. In the stratosphere temperatures are more uniform, but even there we find some vertical changes and, even more so, horizontal changes. The position of the tropopause itself has been marked with a heavy line. We find it above 200 *mb* over Brownsville, Texas. It dips to approximately 250 *mb* north of Dodge City, Kansas, and to below 300 *mb* over Lander, Wyoming. South of each location, where the tropopause dips, a jet finger is located.

In the troposphere we find locations where the smooth course of the isotherms shows a drastic kink. Here air masses of different temperature are lying side by side, separated by a zone of transition. These are called *fronts* or *frontal zones*. In Figures 40–42 their boundaries are marked with heavy lines. We notice that the bulk of the jet stream is located above one such frontal zone, a fact which should be well remembered.

Lines of equal relative humidity have been analyzed in Figure 43. Relative humidity of 100 per cent would indicate complete saturation of air with moisture, and consequently, dense clouds or fog. Zero per cent would signal completely dry air. As it turns out, our radiosonde instruments are not constructed to measure accurately either of these extreme values. Especially under very dry conditions the instrument fails completely, and we have to estimate how much moisture might still be present. Dry regions where the sounding instrument did not perform properly, are outlined by a line labeled *A*. We find

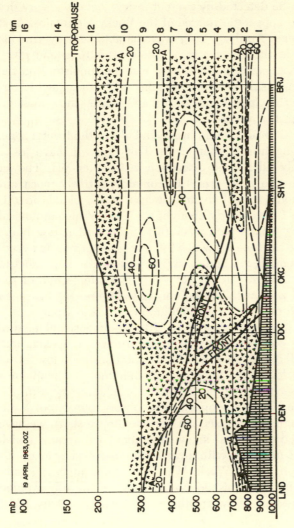

Fig. 43. Lines of equal relative humidity (in per cent) are shown for cross section of Fig. 40, with dry regions shaded.

that generally above 250 *mb* the air has not enough moisture to be detected by present instrumentation. Very dry
air also appears in the lee of the Rocky Mountains. As it
happened, April 19, 1963 was characterized by chinook
winds in the Denver area. These are strong and gusty
downslope winds, which have lost their moisture on the
windward side of the mountains and, consequently, arrive in the Great Plains as dry and warm "snow melters"
and "dust blowers."

Dry air is also found inside and above the frontal zone,
extruding from higher levels. This fact will have to be
remembered for the discussion in Chapter VII. The jet
stream core itself seems to contain relatively moist air.
We should expect to find at least cirrus clouds in this
region. Conveniently enough, the Tiros V meteorological
satellite took a picture of this region just a few hours
prior to the time for which the cross section was analyzed.
Sure enough, we find a wisp of bright clouds parallel to
the jet stream core (Plate VII). The intersection of the
frontal zone with the ground (see Figure 42) has been
entered on this Tiros photograph as a cold front.

We have now developed a three-dimensional picture
of one particular jet stream occurrence. We have analyzed the flow on a quasi-horizontal surface close to jet
stream level. We have cut through the jet stream in a
cross section. We have verified our analyses with the
most modern and sophisticated observational tool available to meteorologists, the spy satellite. In short, we have
acquired peripheral knowledge about jet streams which
puts us on a par with any weatherman trying to do his
job shortly after World War II. Unfortunately, the subject is far from finished. So far we have carefully avoided
the question, Why are there jet streams? Let us ask it
now.

Chapter V

WHAT MAKES A JET STREAM?

A "Mysterious" Deflecting Force

From the observational evidence presented in the previous chapter we have learned that jet streams—and winds in the free atmosphere in general—do not blow directly from high-pressure to low-pressure regions. They are not flowing "downhill" along the fall line of a surface of constant pressure, as a river would flow down a mountain under the force of gravity. Instead, winds have a tendency to blow parallel to the contour lines, circling clockwise around "mountains" of high pressure, and counterclockwise around "valleys" of low pressure in the Northern Hemisphere (see, for instance, Figures 32 and 33). Down Under, in the Southern Hemisphere, the sense of rotation is reversed.

There must be some mysterious force that prevents winds from doing what the rivers do—a force deflecting them from the fall line that points directly from high to low pressure. This line is called the direction of the *horizontal pressure gradient*. The larger the pressure gradient, the steeper the slope from high to low pressure and, consequently, the stronger the pull of gravity. This pull causes winds to be stronger with a strong pressure gradient than with a weak gradient. This we see quite clearly from the maps in Figures 33 and 34. Wherever the contour lines of the 250 *mb* pressure surface are close together, indicating a steep slope of the surface or a large horizontal pressure gradient, the winds are strong. Thus, we find that the jet stream follows the regions of steepest pressure gradients.

From what has been said in Chapter II we recognize the *pressure gradient* as a *vector:* It has a *magnitude* and a *direction* (Figure 44). The latter is normal to the

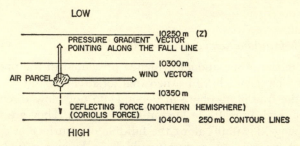

Fig. 44. Vectors representing forces acting on a parcel of air are shown schematically here. Direction of the pressure gradient vector (along fall line from high to low) is upward; wind vector is toward right; vector of the Coriolis force (the deflecting effect of the earth's rotation) is along dotted line.

orientation of the contour lines, and the former is inversely proportional to their spacing. The magnitude of the gradient may be defined as

$$-\frac{\triangle Z}{\triangle n} \qquad (12)$$

where $\triangle Z$ is the height difference of contours, measured along the distance $\triangle n$ on the fall line.

The wind vector blows to the right of the gradient vector in the Northern Hemisphere (to the left in the Southern Hemisphere), due to some "mysterious" deflecting force, shown by a dotted vector in Figure 44. This is the so-called *Coriolis force.* Its presence governs the laws of large-scale atmospheric motions, and thus constitutes a major factor in the formation of jet streams. They, in turn, influence the weather as we experience it in its daily changes (see Chapter VII).

Momentum and the Merry-Go-Round

Have you ever tried to jump off a fast-turning merry-go-round? If you did, you probably remember the bruises on your hands and knees. While you were standing on the outer rim of the rotating platform, you were perfectly comfortable. As you jumped off, however, you took the velocity of rotation with you; therefore, you hit the ground at high speed. Isaac Newton (1687) formulated this experience in a fundamental law of physics: A moving body conserves its *momentum* as long as no outside force acts on it. The momentum, $\vec{M}$, of a body is defined as its mass, m, times its velocity, $\vec{v}$

$$\vec{M} = m \cdot \vec{v} \tag{13}$$

Since velocity is a vector, momentum is a vector pointing in the same direction.

This conservation of momentum makes people fly through the windshields of their cars in a collision, unless they are tied down with seat belts. It makes the ice skater move even after he has stopped pushing with his feet. In a game of pool you try to pass the momentum of the cue ball on to another ball, the vector sum of momentum of both balls being equal to the momentum of the cue ball before collision.

Not only is linear momentum of straight motion, $m \cdot \vec{v}$, conserved if no force from the outside acts on the moving body, but also *angular momentum*. Because angular momentum is conserved, the earth still spins about its axis at approximately the same rate as it did millenniums ago.

Actually, we have to be a little careful with this statement. We may compare the earth to a giant spinning top. A spinning top slows down from the friction of its stem against the ground, and finally topples over. The

earth has no "stem" to "rub" against celestial ball bear-
ings. Nevertheless, there is some friction, and over the
ages it has slowed down the earth's rotation ever so
slightly. Other celestial bodies, especially the sun and
the moon, cause tidal forces of attraction which, in turn,
give rise to motions in the oceans, in the atmosphere, and
in the somewhat plastic crust of the earth. The ocean
tides you see every day if you live on the seashore. The
tidal motions exercise a certain frictional drag on the
earth. This drag is caused by the fact that the two outward bulges of the earth, produced by the moon's force
of attraction, do not lie exactly in line with the moon,
but are carried ahead of the moon by the earth's rotation. Since the moon's attraction on the closer bulge is
stronger than its pull on the bulge facing away, there
will be a net force on these two bulges. This force tries
to pull them back from their rotation ever so slightly.
The pull, or drag, slows down the earth's rotation.

Angular momentum $\vec{G}$ (also a vector, but a more complicated one, as shown in Figure 45) is defined as the
product of the angular velocity, $\vec{\Omega}$, and the moment of
inertia, $m\ r^2$, where m is the mass of the rotating point
under consideration, and r is the radial distance of this
point from the axis of rotation:

$$\vec{G} = \vec{\Omega} \cdot m \cdot r^2 \tag{14}$$

Since $\vec{\Omega}$ indicates a velocity, it is also a vector, pointing
normal to the rotating disk which contains the rotating
point with mass m. According to mathematical conven-
tion the vector points in the direction into which the disk
would "screw in" if it were connected to a "right-handed"
screw. It follows from Equation (14) that the vector G
has the same direction as the vector $\vec{\Omega}$, but different

magnitude, namely multiplied by $m\ r^2$. This comparison is indicated in Figure 45.

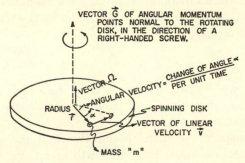

Fig. 45. Angular momentum is a vector perpendicular to the plane of rotation. Its magnitude is $\overrightarrow{\Omega}\ mr^2$, and its direction is that a right-handed screw would take if the screw were fixed to the axis of the spinning disk. $\overrightarrow{G}$ and $\overrightarrow{\Omega}$ have the same direction.

Johannes Kepler (1571–1630) was the first to recognize the principle of Equation (14). He postulated that the radial distance line between a planet and the sun covers equal areas in the orbital plane per unit time, no matter where the planet happens to be in its revolution around the sun. As the planet approaches the sun, its motion speeds up; as the planet recedes, its motion slows down. Ω, the angular velocity, signifies the speed with which the radial distance line (or the radius vector, as we may call it) between sun and planet rotates around the sun. As we remember from geometry, the area covered by such a radius vector in its revolving motion goes with the square of the radius. Thus r^2 gives a measure of the area covered in the orbital plane by a planet revolving at distance r from the sun. Kepler's law expresses the fact that the characteristic product $\Omega \cdot r^2$ of a planet's revolution remains constant. That means the planet has a constant angular momentum.

If you are not familiar with the term *angular velocity*, its definition is very simple: Ω is given by the change of the angle α per unit time, or

$$\Omega = \frac{\alpha}{t} \tag{15}$$

Usually we express the angle α in terms of *radians*, rather than degrees. A full circle of 360 degrees corresponds to 2π radians.*

To compute the angular velocity Ω of the earth is easy, since we know that the earth completes the full circle of 360 degrees in 3 minutes and 57 seconds less than 24 hours (86,400 seconds). Thus,

$$\Omega_{\text{earth}} = \frac{360°}{1 \text{ day} - 237 \text{ seconds}} \tag{16}$$

or

$$\Omega_{\text{earth}} = \frac{2\pi}{86,163} = 7.292 \times 10^{-5} \frac{\text{radian}}{\text{second}} \tag{17}\dagger$$

Equation (14) gives us something novel to consider: Vectors of rotation point in the direction of a screw motion that is at right angles to both the radius vector $\vec{r}$ and the linear velocity vector $\vec{v}$ (Figure 46). This linear velocity has the magnitude of

$$v = r \cdot \Omega \tag{18}$$

According to Equation (18), if we are standing on a disk rotating with constant angular velocity Ω, our linear

---

* Geometrically, the radian is defined as the central angle subtended by the arc of a circle when the arc and the radius are of equal length.

† A factor of 10^{-5} means that 7.292 has to be multiplied by $\frac{1}{100,000}$. Note that the denominator has five zeros.

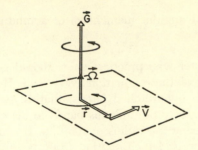

Fig. 46. The direction of angular momentum $\vec{G}$ is normal (perpendicular) to the vector of wind speed $\vec{v}$ and the radius vector $\vec{r}$.

velocity will increase the farther we remove ourselves from the center of rotation; that is, the larger the radius r becomes (compare r_1 and r_2 in Figure 47). If you don't

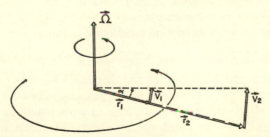

Fig. 47. At constant angular velocity linear speed increases with radius. The body whose linear speed vector is $\vec{V}_2$ travels farther than the body of vector $\vec{V}_1$ does when there is a change in the angle α but the interval of time is the same for both.

believe it, try it out on a Devil's Wheel at the next county fair or carnival you visit.

With the aid of Equation (18) we may express the

magnitude of angular momentum of a rotating body as

$$G = r \cdot v \cdot m \qquad (19)$$

An ice skater who practices spins should get to know the meaning of these equations very well. He starts his rotation with outstretched hands, that is, with a large radius r. Suddenly he pulls his hands to his chest, effectively decreasing the radius of rotation. His angular momentum, expressed by Equation (14), is conserved—except for the slight friction of the skates on the ice—and so is his mass. To compensate for the reduced radius, therefore, the rate of spin, Ω, has to increase drastically. This is the secret trick of the pirouette.

We have compared the earth to a giant spinning top. Actually, we should have been more precise, because this spinning top is made up of three things: the solid earth, the oceans, and the atmosphere. Thus, we should say that the *sum* of the angular momentums of earth, ocean, and atmosphere remains constant (except for the negligibly small effect of tidal friction mentioned before)

$$G_{earth} + G_{oceans} + G_{atmosphere} = \text{constant} \qquad (20)$$

In the previous chapter we discussed the high wind velocities that may occur in jet streams. These atmospheric currents are much stronger in winter than in summer, and there is much more seasonal variation in the strength of jet streams in the Northern than in the Southern Hemisphere. Therefore, the atmosphere as a whole speeds up its mean west-to-east motion during the Northern Hemisphere winter, and slows down during the Northern Hemisphere summer. The oceans behave in a similar manner, but on a much smaller scale, because their circulation essentially is driven by the wind. Thus, if Equation (20) is to balance, the rotation of the earth —contained in G_{earth}—has to decrease as the rotation of atmosphere and oceans increases.

A change in the earth's rate of rotation is felt as an unusual change in the length of a day. The effect is too small to be measured with ordinary watches. One has to measure the length of day (from the passage of a fixed star through the meridian) with atomic clocks because the variations caused by atmosphere and oceans are of the order of milliseconds.

Conservation of angular momentum not only makes the earth go round, it also helps to generate the jet streams. As we will see, the mystery of jet streams may be solved rather easily—in the laboratory we can imitate their birth. But, as so often happens in the so-called simple things of life, the more we discover, the more details we find awaiting explanations. So, even today, we are still far from a complete understanding of all the refined aspects of jet streams.

Let us look at some of the basic facts of the general circulation as suggested in a cross-sectional view of the earth and its atmosphere (Figure 48). The vertical scale in this diagram, of course, has been exaggerated grossly. For a moment we will assume that the earth stands still

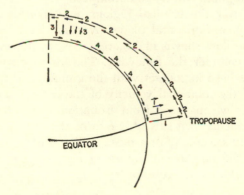

Fig. 48. A cross section of the atmosphere is shown schematically as it would be for a non-rotating earth heated at the equator and cooled at the pole.

and that the sun rotates around as Ptolemy the astrono-
mer made his fellow citizens of Alexandria believe. The
equator of the earth will receive an excessive amount of
heat. This warming will generate convective motions upward to the *tropopause* (arrows "1"). The air thus flowing upward will turn toward the colder pole, as the only
place toward which it can go upon hitting the "lid" of
the tropopause. On its journey it will cool off by radiation
into space (arrows "2"). At the same time, the rising air
over the equator will be replaced by cold air from the
pole flowing along the surface of the earth (arrows "4").
To complete the circulation, air will be sinking over the
poles (arrow "3").

Now let us abandon our geocentric cosmos and look
at things as they really are on a *rotating* earth. We still
must deal with the same thermal conditions. The equatorial regions are receiving more solar energy per square
foot than the poles. To achieve the existing balance of
temperatures—that is, in order to keep the tropics from
frying and the high latitudes from deep-freezing—heat
will have to be transported from the equator to the poles.
The atmosphere will try to accomplish this by convective
motions over the tropical regions, and meridional motions in quasi-horizontal planes over mid-latitudes, similar to the flow shown in Figure 48. Now, however, we
have to consider the following. The air in tropical regions has been in contact with the ground, and thus has
acquired the rotational velocity of the earth at low latitudes. As we may see from Equation (17), the earth
spins about its own axis at a rate of $\Omega = 7.292 \times 10^{-5}$
radian per second. With a mean radius R of 6.371×10^{6}
meters,‡ the *angular momentum of 1 gram of mass* at the
earth's surface at the equator is

‡ The factor 10^{-5} means that we have to divide by 100,000
(which has five zeros). A factor of 10^{6} means we have to multiply
by 1,000,000 (six zeros).

$$G_1 = (6.371 \times 10^6)^2 \times 7.292 \times 10^{-5} \times 1$$

$$= 295.980 \times 10^7 \ m^2 \cdot \text{radians} \cdot sec^{-1} \quad (21)$$

(For simplicity we will henceforth consider angular momentum per unit of mass; that is, we will assume that in Equation (14) $m = 1$.)

If an air parcel has acquired the earth's rotational velocity at the equator—that is, if the parcel rotates at the same speed as the earth—an observer in the tropics will report this as "no wind." He, of course, also rotates with the earth, together with the air parcel, and he will feel no breeze.

As the air parcel, which has risen over the equator to the tropopause, moves poleward, several things start happening. First of all, with increasing latitude ϕ the distance r from the axis of rotation decreases because of the curvature of the earth. The dependence of r on latitude may be seen from Figure 49 and may be expressed

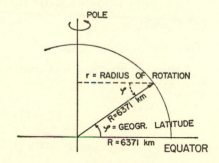

Fig. 49. The radius of rotation $\vec{r}$ decreases with geographic latitude.

in simple trigonometric terms

$$r = R \cdot \cos \phi \quad (22)$$

If the air parcel conserves its angular momentum on the trip—and this it will do as long as no force acts on it

—its rate of rotation relative to the earth's axis will have to change in order to compensate for the shrinking radius

$$G = (R \cdot \cos \phi)^2 \cdot \Omega = \text{constant} \qquad (23)$$

Let us watch the parcel as it travels from the equator ($\phi_1 = 0°$) to latitude $\phi_2 = 30°$. We assumed that the parcel had the earth's rate of rotation at the equator, meaning that there was no wind where the parcel started rising to the tropopause. Thus,

$$G_1 = (R \cdot \cos \phi_0)^2 \cdot \Omega_1 = R^2 \cdot \Omega_1 \qquad (24)$$

because $\cos 0° = 1$ and $\Omega_1 = \Omega = 7.292 \times 10^{-5}$ radian $\cdot sec^{-1}$, which is the rotational velocity of the earth. From Equation (21) we see that $G_1 = 295.980 \times 10^7 \ m^2 \cdot$ radians $\cdot sec^{-1}$.

At latitude 30° the parcel will have the angular momentum

$$G_2 = (R \cdot \cos \phi_2)^2 \cdot \Omega_2$$
$$= (6.371 \times 10^6 \times 0.866)^2 \ \Omega_2 \qquad (25)$$

because $\cos 30° = 0.866$, or

$$G_2 = 30.440 \times 10^{12} \times \Omega_2 \qquad (26)$$

Now, if the angular momentum was conserved during this trip, we have to write

$$G_1 = G_2 \qquad (27)$$

or

$$295.980 \times 10^7 = 30.440 \times 10^{12} \times \Omega_2$$

From this we may calculate the rate of rotation of the air parcel relative to the earth's axis

$$\Omega_2 = \frac{295.980 \times 10^7}{30.440 \times 10^{12}} = 9.723 \times 10^{-5} \text{ radian} \cdot sec^{-1} \quad (28)$$

We see that the air parcel now, as it arrives at 30° latitude, rotates considerably faster than the earth itself [see Equation (17)]. Since the earth turns from west to east, the higher rate of rotation of the air will show as a strong *west wind.*

We usually consider the linear velocity of winds, not their rate of rotation. The conversion is easy with the help of Equation (18). We have to remember, however, that as observers standing on the earth we are interested only in the speed of the air *relative* to the earth, so we have to subtract the angular velocity of the earth from the rate of rotation of the air computed in Equation (28).

Thus, the wind speed, in familiar units of *mps*, is

$$u = (R \cdot \cos 30°) \times (\Omega_2 - \Omega) \text{ or} \qquad (29)$$

$$u = (6.371 \times 10^6 \times 0.866) \times [(9.723 - 7.292) \times 10^{-5}]$$
$$= 134.1 \text{ } mps \qquad (30)$$

Other Forces

From this brief example we see that under conservation of absolute angular momentum relatively small displacements of air poleward would suffice to generate winds of jet stream velocity. From the examples in the previous chapter we see that actual wind speeds in the jet stream are less than the one just computed. This disparity indicates that angular momentum is *not* strictly conserved during the poleward journey of equatorial air. Some external force or forces must be acting on the air parcels while they travel.

The most important of these are the so-called *pressure forces.* The air parcels which have risen in convective motions and cloudy weather over the tropics and are now moving poleward will have to push against air occupying their path. There will be a certain piling up—or *convergence,* as the technical terminology reads—of air. This surplus of air will generate a high-pressure ring around the whole hemisphere, and tend to force the flow outward. Nevertheless, the equatorial air continues to pile up against this region of high pressure, battling the pressure forces. These forces prevent the air from going farther north, and from attaining still higher velocities under conservation of angular momentum.

We have reasoned that the convergence aloft, within the flow from the equator, coincides with increasing westerly wind speeds which try to conserve their angular momentum. Convergence, however, keeps piling mass into a vertical air column, thus increasing its total weight from the top of the atmosphere to the ground. Since the surface pressure, measured by a barometer, is nothing but this total weight of the air column, the barometric pressure will rise as air piles into the vertical column at jet stream level. Thus, convergence of air generates a high-pressure belt underneath the belt of strong west winds aloft. Some of this air tries to escape vertically from this high-pressure region, but the path upward is blocked because the stable stratification of the stratosphere presents another force to fight against. Most of the vertical flow, thus, is directed *downward*. Sinking air is subjected to higher atmospheric pressures, and, therefore, will be warmed. Clouds, if they exist, should evaporate. Thus, we can expect fair weather in this high-pressure belt (Figures 50 and 51).

Near the equator the air rises. It cools in rising because it expands into a low-pressure environment. Moisture will condense and form clouds and precipitation. Hence, tropical regions have abundant rainfall.

At the earth's surface the air will be flowing out, or diverging, from the sub-tropical high-pressure region. Part of it will continue toward higher latitudes [(1) in Figure 51]; part, however, will return toward the equator [(2) in Figure 51]. Not only will this air lose its westerly speed by friction in the lower atmosphere and at the ground; its flow toward the equator will let us apply Equations (24) through (30) "backward." We now start at latitude ϕ_2 and end up at latitude ϕ_1, ob-

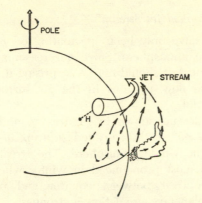

Fig. 50. Schematic view gives a three-dimensional picture of air movement in a rotating earth heated at equator and cooled at high latitudes.

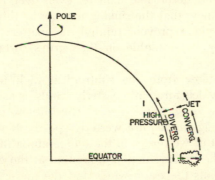

Fig. 51. A plane cross section of the situation in Fig. 50.

taining a negative speed *u* of the same magnitude as before. The minus sign signifies an *east* wind.

Among other things, these easterlies will have to fight constantly against friction at the earth's surface. This battle consumes most of their energy, making them rather weak, very much weaker than would correspond to a southward flow under conservation of angular momentum.

The "Subtropical Jet Stream"

So far we have only been speculating on *possible* motions in the atmosphere. Since our reasoning has been entirely logical, it should not be surprising if we actually found these flow patterns in the air surrounding our globe. We do!

The jet stream shown in Figures 50 and 51 at approximately 30° is the so-called *subtropical jet stream.* It circles the whole hemisphere. The stream is rather steady and most pronounced in winter when the temperature contrast between equator and pole, which drives the planetary circulation, is strongest.

Underneath the subtropical jet we find (as we postulated) the *subtropical high-pressure ridge* with fair weather and desert-like climate, even over the oceans. We now know that the sinking motions in this region of the atmosphere prevent rainfall, no matter how much water may be available on the earth's surface underneath.

The outflow from the subtropical high toward the equator, which turns into easterly winds, is easily identified as the *trade wind region.* The outflow toward the pole will, again, acquire westerly speeds. It contributes to the belt of westerlies in the temperate latitudes (Figure 1). The braking action of friction at the earth's surface prevents the trade winds and the mid-latitude surface westerlies from attaining jet stream velocities.

With the foregoing arguments we come to recognize the whole mechanism of the tropical and subtropical circulation of the earth's atmosphere in a new, but physically realistic, concept. The trade winds are part of a meridional circulation system that also contains as an important part the subtropical jet stream and its convergence of air aloft. The system, in turn, is maintained by the outflow of equatorial air near the tropopause level under quasi-conservation of angular momentum. It is all as simple as that.

But in all its fascinating logic and simplicity the atmosphere still has a few tricks in store for meteorologists. If we wanted, for instance, to compute and forecast the magnitude and variations of the outflow from the equator, which generates the subtropical jet stream, we still would have to contend with pitifully inadequate observational data. This gap, however, is closing fast, and the not-too-distant future may see scores of balloons drifting majestically at lofty heights, sending their measurements of temperature, pressure, and location by thin-film electronic transmitters, which are "printed" on their skin, to monitoring satellites. The flight of such balloons will map the winds in the skies over regions which are claimed neither by birds nor by planes, yet which may harbor the nucleus for future weather development over far distant regions of the globe.

Jet Streams in the "Dishpan"

We might ask a question. Why do we find the subtropical jet near 30° latitude? Why does the meridional circulation not extend all the way to the pole? You can provide the answer yourself. According to Equation (24), the angular momentum of a unit mass of air, as it leaves the equator, is $G_1 = R^2 \cdot \Omega = 295.980 \times 10^7 \ m^2 \cdot$ radians $\cdot sec^{-1}$. If this air parcel traveled all the way to the *pole* under conservation of angular momentum, according to Equations (25) and (26) its rotational conditions upon arrival would be given by $G_1 = G_2 = (R \cdot \cos \phi_2)^2 \cdot \Omega_2$. Since at the pole $\phi_2 = 90°$ and consequently $\cos \phi_2 = 0$, $(R \cdot \cos \phi_2)^2$ is also zero. In order to fulfill the condition $295.980 \times 10^7 = 0 \times \Omega_2$ the value of Ω_2 would have to become infinitely large. A gale of unimaginably high velocity would have to blow at the pole. Moreover, since this fantastically strong jet stream would have to circle the pole, winds a small distance from the pole would be blowing from opposite directions. In other words, over each of the earth's poles a monstrous tornado would rage constantly. The horizontal wind shear would be terrific and would immediately cause turbulence. The whole jet stream would have to break down into eddies of varying size.

We may watch this process in laboratory experiments, the so-called "dishpan experiments." A dishpan filled with water is rotated on a shaft. A copper cylinder in the center of the pan is supplied with iced water, representing the "cold pole." Along the outer edge of the bottom runs a coil of heating wire, simulating the "hot equator" (Figure 52). Although the dishpan is flat, it

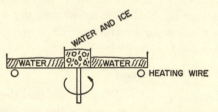

Fig. 52. The rotating "dishpan" for experiments simulating atmospheric conditions is shown schematically in cross section.

has the following important characteristics in common with the earth's atmosphere: (1) The temperature difference between "equator" and "pole" necessitates a meridional transport of heat; (2) since the whole system is rotating, angular momentum will have to be transported together with heat. This combination will give rise to jet streams.

We may start the experiment with very slow rotation, much slower than the earth's rate of spin (allowing for the vast difference in dimension). First, we observe a symmetric jet stream of moderate intensity close to the polar cooling cylinder. As we increase the rate of rotation, this jet stream becomes stronger and stronger. At the same time, it moves away from the pole, because frictional forces prevent the "wind shear" near the cooling cylinder from becoming too large. Still the jet stream is perfectly symmetric about the pole. A meridional circulation cell similar to the one shown in Figure 51 but reaching all the way from the equator to the pole maintains it.

With a further increase in the rate of rotation, the symmetric jet stream suddenly breaks down into a pattern of symmetric waves, first two, then three, four, and so on, depending on the speed of rotation. Up to seven waves have been observed. This breakdown indicates that the meridional circulation is not capable any longer of efficiently transporting heat and momentum poleward. Horizontal eddies have to help in this transport. They do, by transporting warm water in the southwesterly flow and cold water in the northwesterly current.

These horizontal eddies in the dishpan are the counterparts of large cyclones and anticyclones in the atmosphere. We come to realize now how important the storms and blizzards over the midwestern United States are. They may cause considerable hardship to man and beast, yet at the same time they carry the burden in the transport of heat and momentum. Without this transport the atmosphere could not maintain its climatic balance over the years. Therefore, when our heating bill goes up in the blizzard season, or when the air conditioners have to go full blast while a summer anticyclone has settled itself comfortably over our country, let's be grateful for these hardships because they help to make our continent habitable.

Plate VIII is a photograph of a symmetric three-wave jet stream in a dishpan. Such photographs are obtained by so mounting a camera that it aims down at the pan

and rotates at the same speed. The camera, thus, is like
an observer standing on the rotating earth. It sees the
relative motion of the water in the dishpan, as an observer senses the winds *relative* to the moving earth.
Aluminum powder scattered on the surface makes the
water motions visible. A time exposure will produce a
streak on the film for each moving particle. If a flashbulb
is exploded at the end of the time exposure, a bright dot
will appear at the end of each streak. The length of a
streak indicates the speed at which the water moves in
the pan; the dot marks the direction of motion.

If we speed the rate of rotation up further, the symmetric waves no longer will be sufficient for the transport
of heat and momentum. Strongly "tilting" troughs develop in the jet stream, sloping from northeast to southwest (Figure 53 and Plate IX). They amplify until they

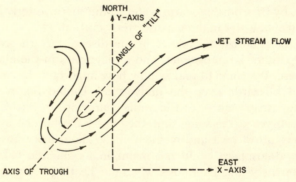

Fig. 53. The "tilting trough," whose flow pattern is shown
schematically here, develops in jet stream when heat transport system breaks down.

break down into *cut-off* cyclonic and anticyclonic vortices, separated from a newly developing jet current,
which undergoes the same development (Figure 54).

This unstable state of the dishpan circulation resem-

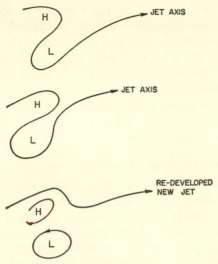

Fig. 54. "Cut-off" cyclones and anticyclones develop from jet stream around high-index (H) and low-index (L) flow patterns as suggested here.

bles closely what we observe in the atmosphere. Here, too, the jet stream systems undergo certain phases of development. *High-index* stages of flow pattern, in which the jet streams are oriented almost zonally (that is, nearly parallel to latitude circles) alternate with *low-index* patterns, in which low-pressure troughs and high-pressure ridges of large meridional amplitude prevail. Obviously, the earth's atmosphere does not manage to transport the necessary amounts of heat and momentum by either a symmetric jet stream or a symmetric wave pattern. If it did, all weather forecasters would be out of their jobs. The weather would follow such a regular schedule that you could plan your picnic a year ahead.

The tilting troughs and ridges shown in Figure 53 are very efficient in transporting momentum of westerly winds poleward. Ahead of the trough southwesterly

winds prevail, indicating that a southerly wind component carries westerly momentum northward to the pole.
To the rear of the tilting trough winds blow from the
northeast. There a northerly wind component carries
easterly momentum southward. In mathematical and
physical terms this situation is equivalent to transporting
westerly momentum northward. Thus, both sides of the
tilting trough help in accomplishing the necessary poleward transport of momentum.

"Eddy" versus "Mean" Transport

Let us express these considerations mathematically. In
first approximation we may consider the earth as a flat
plane, neglecting its curvature. We may, in this case, describe the motions of the atmosphere in a rectangular
coordinate system, whose positive x-axis points eastward,
and whose positive y-axis is directed toward the North
Pole (Figure 53). We may now express the wind vector
$\overrightarrow{v}$, measured in a jet stream, in terms of its west- and
south-wind components, u and v (Figure 55).

Fig. 55. Wind vector $\overrightarrow{V}$ can be resolved into its components,
u and v. U is the west wind component, v the south wind
component.

The westerly momentum of an air parcel is given as

$$m \cdot u \qquad\qquad (31)$$

where m is the mass of this parcel and u is its westerly
velocity component. The rate of poleward transport of
this westerly momentum may be computed by multiply-

ing the momentum $m \cdot u$ with the speed v with which it is carried northward. Thus, we have

$$m \cdot u \cdot v \qquad\qquad (32)$$

for the poleward momentum transport.§

An east wind has a negative u-component of motion, as it blows in the direction of the negative x-axis. By the same token, a north wind has a negative v-component. The transport of momentum associated with the northeast winds to the rear of the trough shown in Figure 53 is

$$[m \cdot (-u)] \cdot (-v) = m \cdot u \cdot v \qquad (33)$$

quantity to be transported

velocity of transport

Thus, if westerly, or angular momentum is to be transported northward, v-components of motion have to be present in the general circulation of the atmosphere.

§ The *rate of transport*, also called *flux*, gives the velocity with which a certain property is transported into a certain direction. Instead of westerly momentum $(m \cdot u)$ we may want to consider people on a ski lift. Let us assume that the chair lift has a capacity of two persons per 50 feet of cable. Instead of the rate v of *northward* transport of momentum, we consider the speed c with which the cable moves up the hill, let us say 5 *fps*.

The rate at which the skiers will be deposited on top of the hill conforms to the rate of transport, or the flux, of people along the chair lift. It may be computed as

$$\left(\frac{2 \text{ persons}}{50 \text{ feet}} \right) \times 5 \ \frac{ft}{sec} = \frac{1 \text{ person}}{5 \text{ second}}$$

If we want to transport westerly momentum instead of ski bunnies, we follow an analogous computation

$$(m \cdot u) \times v = \text{northward momentum flux, or rate of transport}$$

quantity to be transported

velocity of transport

From the preceding discussion we have seen that such
v-components may materialize in two different ways:

(1) By *sheets* of air moving poleward near the tropopause and equatorward near the ground. Such motions
could be maintained in the tropical and subtropical regions all around the globe, generating a huge meridional
circulation *wheel,* as shown in Figure 51. Even if we
were to consider only the mean motion in a meridional
plane, averaged over all geographic longitude sectors, we
would still find the presence of a positive v-component
near the tropopause, and of a negative v-component
near the ground. As we have seen, this mean motion
represents a very powerful and efficient mechanism in
generating jet streams through the tendency to conserve
angular momentum. Since this transport is present even
in a mean meridional cross section through the atmosphere, we call it the momentum transport by the *mean
meridional circulation.*

(2) At the earth's rate of rotation, this mean meridional circulation does not suffice any longer to transport
enough momentum and heat to middle and high latitudes
in order to maintain a climatic balance in the atmosphere.
Consequently, the smooth circulation wheel breaks
down, and troughs and ridges develop in the jet stream
flow. They are associated with cyclones and anticyclones. If we traveled around the globe at these latitudes,
we would find alternating regions with northward and
southward flow, or with positive and negative v-compo-
nents of motion. Adding all these components, and
computing the mean meridional motion, we would see
that the latter is almost zero, since the positive and
negative v-components have a tendency to cancel each
other when averaged around the globe. Thus, the mean
flow transports hardly any momentum poleward (see,
for instance, Figure 38). However, since quite frequently
large values of u occur with positive values of v (strong

southwest winds) and small, or even negative, values of
u appear when v is negative in these troughs and ridges
especially when they are tilted (weak northwest or
northeast winds), the product "$u \cdot v$," added around the
globe, will be strongly positive. We may consider these
meanders in the jet stream flow as large eddies in the
circulation of the atmosphere, measuring several thousand kilometers across. These are typical dimensions of
very large cyclones and anticyclones. The transport of
momentum effected by these eddies is called the *eddy
transport.*

Thus, we may say that both forms of transport—the
mean meridional circulation transport and the *eddy
transport*—together perform the task of moving angular
momentum and heat poleward. At tropical latitudes the
former is more efficient; at middle and high latitudes the
latter clearly dominates. Both transport forms maintain
the large-scale motions of the atmosphere, and they feed
the jet streams with energy.

Chapter VI

PANDORA'S BOX

The professional life of a meteorologist would be rather simple if all discussion of the jet streams could end right here. We would have only one jet stream to contend with, and we could explain how it is maintained by the tendency to conserve angular momentum. The exciting discoveries of the World War II bomber pilots could be explained theoretically, and could be verified or demonstrated in the laboratory with the dishpan experiments we have described.

Unfortunately, however, the atmosphere does not cooperate in such simple and clear-cut solutions. No sooner was the jet stream born, than meteorologists began to realize that the atmosphere contains a variety of wind systems which all qualify as jet streams according to the definition by the World Meteorological Organization, given in Chapter I.

The higher our measurement probes, such as balloons and rockets, reached into the atmosphere, and the more accurate and sophisticated our measurements became, the more details and problems surrounding the jet streams were uncovered. Today, two decades past the discovery of high-speed rivers of air over Japan and the Mediterranean, research still continues to explore the secrets of atmospheric flow patterns.

Jet Streams as a "Pause" Phenomenon

In trying to bring some semblance of order into the variety of jet stream systems, we will make use of the various layers and pauses in atmospheric structure, described in Chapter I. Let us start with the *troposphere* and its upper lid, the *tropopause*. As has been mentioned, the tropopause is approximately 16 *km* high over the equator, and only about 8 *km* over the pole. Tropospheric temperatures are high over the equator (surface temperatures ranging close to thirty degrees of the Centigrade scale in the yearly average), and low over the poles (considerably below freezing). The low stratosphere, however, is cold over the equator (mainly because it is located at such high levels, and, as we know, temperature decreases with altitude in the troposphere), and warmer over the pole (considerably warmer during summer, when the atmosphere in these latitudes receives solar radiation twenty-four hours a day).

Cold air is denser than warm air; it is, therefore, heavier, and pressure decreases more rapidly as we go up in the atmosphere. That this is so can be seen easily if we compare conditions in two glasses filled to the same height, one with water, the other with mercury. Pressure is no more than the total weight of the column of liquid (or air) per unit area of the bottom surface. On top of the liquid the pressure is zero* in both glasses. If we remember that *density* is the mass of one cubic centimeter of the liquid and that *weight* is mass times the acceleration of gravity, we find that pressure is given as

Pressure = height (in centimeters) × density × gravity

or

$$P = h \times \rho \times g \qquad (34)$$

In meteorology it is customary to use the Greek letter ρ (rho) as a symbol for density. For water density is

---

* Actually it is not zero, but equal to the pressure of the atmosphere resting on both glasses. If the two glasses are standing next to each other, the atmospheric pressure will be the same in both cases; hence its effect may be neglected for the purpose of the following discussion.

roughly equal to 1. In our latitudes, gravity is approximately 980 cm/sec^2. Thus, on the bottom of a tumbler filled with 10 cm of water, the pressure will be

$$P = 10 \ cm \times \frac{1 \ g}{cm^3} \times 980 \ cm/sec^2 \qquad (35)$$

or 9800 $g/cm \ sec^2$.

The density of mercury is approximately 13.6 g/cm^3. Thus, the pressure at the bottom of the glass will be

$$P = 10 \ cm \times \frac{13.6 \ g}{cm^3} \times 980 \ cm/sec^2 \qquad (36)$$

or 133, 280 $g/cm \ sec^2$.

Since the pressure at the bottom of the glass of mercury is 13.6 times higher than at the bottom of the water glass while the pressure on top of both glasses is the same, clearly the pressure has to decrease faster with height in the mercury glass than in the water glass.

Density of air near sea level at 0° C is approximately 0.0013 g/cm^3, but at 30° C it is approximately 0.00115 g/cm^3. The difference is not nearly as great as between mercury and water, but nevertheless it is there.

With what we have deduced so far, we could sketch a schematic cross section from pole to equator, showing hypothetical pressure surfaces (Figure 56). At the earth's surface we have high-pressure (or anticyclonic) conditions over the pole and low-pressure conditions over the equator, conforming to the distribution of cold and warm temperatures.† This contrast agrees with the fact that cold air is denser and, therefore, weighs more per cubic foot than warm air.

Since pressure decreases faster vertically in cold air than in warm, successive constant pressure surfaces will be closer packed in the former than in the latter.

† We will neglect for a moment the "piling up" of mass described on page 103, which generates a high-pressure ridge in subtropical latitudes.

Whereas near the ground the pressure surfaces slope from north to south (Figure 56) due to the difference in

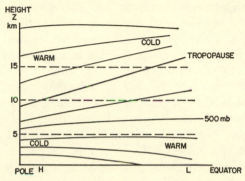

Fig. 56. Surface pressure conditions may reverse with altitude, as shown here, because in cold air pressure decreases more rapidly with height than it does in warm air.

vertical packing somewhere in the middle troposphere, this slope is reversed from south to north. At a height of 5 *km*, for instance, we already find lower pressure values over the pole (where we had an anticyclone near the ground) than over the equator. Correspondingly, the 500 *mb* surface will be found at lower elevations over the pole than over the equator. Thus, we are able to explain why "lows" in the upper troposphere are *cold,* and "highs" are warm.

The steepness in the south-to-north slope increases until we reach the tropopause. Above the tropopause we find a reversal of temperature distribution; warm over the pole, cold over the equator. This reversal is indicated to a certain extent by the 250 *mb* temperatures shown in Figure 35. Correspondingly, the slope of the constant pressure surfaces will gradually decrease above the tropopause, at least during summer conditions when the polar stratosphere is heated by the sun. Why? In a warm

stratosphere pressure surfaces are less closely packed in
the vertical than they are in a cold stratosphere.

We remarked earlier that winds do not simply flow
from high-pressure regions into low-pressure regions.
They try, but the earth's rotation deflects them—toward
the right in the Northern Hemisphere, toward the left
in the Southern Hemisphere. Ultimately, when equilib-
rium between pressure gradient force and deflecting
force is reached, the winds will flow parallel to a
constant-height (or contour) line circling the lows counterclockwise in the Northern Hemisphere, and the highs
clockwise. (In the Southern Hemisphere, the sense of
rotation is reversed.) These winds, flowing parallel to
contour lines, are called *geostrophic winds* (page 68).

Still, the driving forces for these winds are the horizontal pressure gradients; the steeper the slope of the
pressure surfaces, between high and low, the faster the
winds try to "rush downhill," and the faster they are de-
flected and forced into a path parallel to contour lines
(compare Figures 33 and 34).

Returning to Figure 56, we find that near the pole and
close to the ground the air motions tend to follow the
southward slope of the pressure surfaces. Deflection toward the right thus generates east winds at high latitudes
near the earth's surface.

At a level where the pressure surfaces run horizontal,
there will be no wind at all. As we go higher in the
atmosphere, the south-to-north slope of the pressure surfaces will give rise to west winds. As the slope increases
with height, so will the speed of these westerlies. Maximum slope is reached near the tropopause. Maximum
west winds, therefore, will be found near tropopause
level. We may define in good approximation the tropopause as a level at which jet streams may occur. These
we may classify as *tropopause jet streams*.

Furthermore, the poleward transport of angular momentum, described in the previous chapter, is strongest near the tropopause (Figure 48). The jet stream driven and maintained by the tendency to conservation of angular momentum will for this reason, too, be located near tropopause level.

We have mentioned that things are not as simple in the atmosphere because we have pressure forces, besides conservation of angular momentum, to contend with. Such pressure forces are generated by the juxtaposition of cold and warm air masses. If the juxtaposition occurs along a longitude circle, and not along meridians only, it will mean that troughs and ridges are present in the pressure distribution, giving rise to eddy motions of the jet stream. These have been discussed in the previous chapter.

Furthermore, the temperature does not change gradually between equator and pole (as assumed in Figure 56) but most of the contrast between air from polar and tropical sources is concentrated in narrow zones, the so-called *fronts*.

With such a sharp contrast in temperatures, the pressure distribution will also change more abruptly than indicated in Figure 56. Most of the horizontal pressure gradient will be concentrated in and above the frontal zone, where the difference in temperature prescribes the difference in vertical spacing between constant-pressure surfaces. This condition is shown in an exaggerated way in Figure 57. Since the slope of these pressure surfaces determines the strength of the geostrophic wind, we have to expect *jet stream winds over each frontal zone*, attaining their maximum velocity near tropopause level. The cross sections shown in Figures 40 and 42 bear this supposition out clearly.

If the cold air over the pole formed a symmetric bubble, separated from the tropical air by a frontal surface surrounding the globe, we should expect to find a sym-

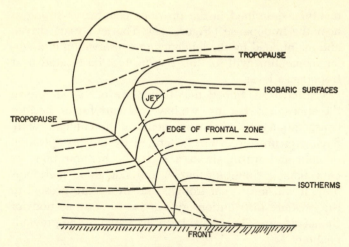

Fig. 57. Across a frontal zone the sharp temperature contrast causes pressure surfaces to slope strongly. A relatively narrow band of strong winds develops aloft—the jet stream.

metric jet stream at tropopause level, blowing around the earth (Figure 58). This jet stream would receive its

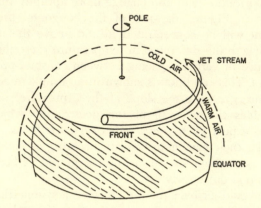

Fig. 58. The jet stream flows over the boundary between cold and warm air.

energy from the pressure forces acting between cold and
warm air. Specifically, cold polar air tries to sink in underneath warm tropical air, the latter trying to rise over
the former. This movement would lower the common
center of gravity in both air masses, liberating *potential
energy*. (This is the energy stored by a certain mass ele-
vated at a certain height. The potential energy of water
in a storage reservoir in the mountains provides a power
plant in the valley with a rushing torrent of kinetic energy, when the mass of water at level of the reservoir
descends to the level of the power plant.)

"Subtropical" and "Polar Front" Jets

In reality, the border between cold and warm air is
not symmetric, but meanders around the hemisphere in
low-pressure troughs and high-pressure ridges. So does
the jet stream accompanying the undulating frontal zone.
This wavelike motion also appears in dishpan experiments, if the rate of rotation is sufficiently high. This we
have discussed in the previous chapter. The "waves"
themselves are a manifestation of large cyclones and anticyclones, as we see them on a weather map. Since these
meanders vary in position from day to day, they will be
obliterated if we consider *mean* flow conditions for a
month or a whole season. By this averaging we arrive at
the cross section shown in Figure 59.

Now we find *two* westerly jet streams at tropopause
level. Both will qualify as tropopause jet streams. The
southern one is maintained by transport of angular momentum in the rising convective currents over the tropics, which spread northward in low latitudes. Underneath
this jet (as explained in Chapter V) we find the *subtropical high-pressure belt* with air currents diverging at
the earth's surface and thereby dissolving any fronts. This
jet stream is, therefore, called the *subtropical jet stream*.
Its strongest winds are found near a height of 13 to 14

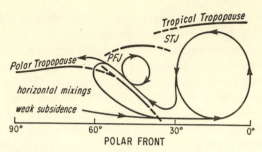

Fig. 59. Meridional circulation cells and the two main jet streams (the polar front and the subtropical) are shown schematically here in cross section. (After E. Palmèn)

km. There is no front near the ground associated with this jet stream. Winds may reach more than 250 knots, especially over Japan, where our bomber pilots ran into trouble.

The second jet stream in Figure 59 is found at slightly lower levels, near 10 *km.* It flows over a well-developed frontal zone, the so-called *polar front.* Its name, therefore, is *polar front jet.* It is this jet stream which breeds stormy weather and blizzards over the United States.

If we wish to be very specific and detailed, we may mention that the jump between equatorial and polar temperatures may be concentrated into several instead of only one frontal zone. Each of these fronts would carry its own jet stream. Thus, meteorologists may distinguish between *Arctic front jet streams* (marking the front between very cold Arctic air and moderately cold polar air) and *polar front jet streams* (associated with the front between polar and tropical air). In origin, nature, and behavior they are very much alike.

Something about Energy

At this point, let us review the important place of these two jet streams in the atmosphere's general circulation.

The driving agent of this circulation is solar energy received mainly at the earth's surface in low latitudes. For a climatic balance to continue over the centuries, the same amount of energy must be lost by the atmosphere to outer space. The excess of heat loss over incoming radiation is greatest in high latitudes. The latitudinal differences in radiation balance lead to a temperature distribution similar to the one sketched in Figure 56; warm troposphere over the equator, cold troposphere over the pole.

This temperature gradient, directed from equator to pole, represents a reservoir of *potential energy,* because cold air has a tendency to slide in underneath warm air, thus lowering the common center of gravity for the whole atmosphere. So far, we have assumed tacitly that the radiation processes are distributed symmetrically about one hemisphere (either the Northern or the Southern). Hence, the buildup of potential energy resulting from these processes should also be symmetric about each pole. It will reveal itself very distinctly if we average temperatures and pressures around each circle of latitude, thus arriving at a mean distribution of potential energy. We, therefore, call this symmetrically distributed form of energy the *mean potential energy* (MPE).

This MPE would tend to produce a meridional circulation (Chapter V), thus activating the subtropical jet stream. From dishpan experiments we know that the original tendency of the meridional circulation is to produce a jet stream belt which is axially symmetric to the pole of rotation. Again, this motion of air around the earth's axis would be brought to light very clearly by averaging the atmospheric wind speeds around circles of latitude. We, therefore, call this the *mean kinetic energy.*

From all the foregoing discussion we see that conversion of mean potential energy into mean kinetic energy

has an important effect in the maintenance of the subtropical jet stream (Figure 60). From our reasoning in

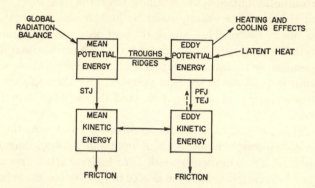

Fig. 60. The energy cycle in the atmosphere. STJ denotes
subtropical jet stream; PFJ, polar front jet; TEJ, tropical
easterly jet.

Chapter V we may furthermore deduce that the rising
warm tropical air (which releases the mean potential energy of the whole circulation system and thereby generates mean kinetic energy) tends to conserve angular
momentum as it moves poleward and gains strong westerly wind speeds.

    This may all sound very confusing now. What actually
generates the subtropical jet stream? Is it conservation
of angular momentum, or is it release of mean potential
energy? Both processes, shall we say, have to be satisfied
simultaneously. We may compare our dilemma with two
persons visiting a museum and viewing a statue. One person looks at it from the front, the other from the back.
Both see the same statue and understand its meaning
and purpose (unless the statue is of contemporary de-
sign), although each of the viewers sees something al-
together different.

    The angular momentum aspect of the jet stream exhib-

its the atmosphere as a closed system with no external
moving forces, which has to conserve its momentum of
rotation as a whole. *Inside* this system, however, friction
at the ground generates and destroys momentum. (East
winds, for instance, lose easterly momentum by the
braking action of friction, and thereby gain westerly momentum.) Therefore, angular momentum will have to be
transported through the atmosphere in order to balance
the gains and losses near the ground. It is these transports which we used in Chapter V to explain the jet
stream.

At the same time we may argue that motion has to be
generated by forces issuing from a reservoir of energy.
The total energy of the atmosphere, however, should re-
main constant if we are not to experience drastic changes
in climate. We may state that

Total energy = potential energy + kinetic energy + heat
(or internal) energy

From this equality we see that if total energy is conserved, a decrease in potential energy will lead to an
increase in kinetic energy. (Since *internal energy* bears
a fixed ratio to potential energy, we will not mention it
separately any longer but consider it included in the ex-
pression "potential energy.") Thus, we may explain the
existence of jet streams by reference to energy conversions, without violating the angular momentum concept.
We are just looking at two sides of the same phenomenon, the jet stream.

Let us now return to Figure 60. At the earth's present
rate of rotation and under the temperature gradient existing between equator and pole, a simple symmetric jet
stream band cannot be maintained, we know, because
such a band does not transport enough momentum and
heat. The circulation breaks down into meanders and eddies. In addition to the simple temperature distribution

of Figure 56—warm in the tropics, cold at the poles—we have a side-by-side sequence of pockets with cold and warm air as we go around the globe in middle latitudes. The cold air pockets may be associated with cold high-pressure regions at the surface. But aloft, because of the greater density of cold air and the closer vertical spacing of constant-pressure surfaces therein, we will meet cold low-pressure troughs at jet stream level. By the same token, high-pressure ridges aloft are warm (see the analyses of 250 *mb* charts shown in Chapter VI). Cyclones and anticyclones at the surface will be associated with the larger ones of these troughs and ridges aloft.

Since we are now faced with eddies of warm and cold air resting side by side along circles of latitude, we call the potential energy resulting from this distribution the *eddy potential energy.*

Again, cold air will try to slip in underneath warm air, the latter trying to rise over the former. This is typically the case along fronts of mid-latitude cyclones: Cold air pushes in underneath warm air, behind the cold front. Frontal clouds in this region result from cooling and subsequent condensation processes in the warm air which is pushed upward by the wedge of cold air. Over the warm front, warm air glides up over a wedge of slow-moving cool air. In doing so, it cools and (Figure 61) pro-

Fig. 61. Frontal system associated with cyclone is represented here in cross section.

duces clouds and precipitation. (We will learn more about cyclones a little later.)

The release of potential energy, resulting from the sinking of cold air and the rising of warm air, will serve to produce kinetic energy. The latter will not necessarily be tied to motions in a west-to-east direction; there will also be south winds and north winds generated in this energy conversion. Thus, we see *eddy kinetic energy* produced in the atmosphere.

Again, we may tie this reasoning in with earlier discussions. As we have seen, these large eddies in the jet stream circulation help to transport more angular momentum poleward (especially when the trough lines tilt from the southwest to the northeast), than a mean meridional circulation can accomplish. This line of reasoning supports the argument that a momentum balance must be achieved if we are to consider the atmosphere as a closed system. Our description and discussion of the energy conversions start, of course, from the premise that the total energy of the atmosphere is conserved.

By heating and cooling processes the meandering distribution of cold and warm eddies may lose some of its effectiveness. Cold air moving equatorward in a trough has a tendency to warm as it crosses warmer terrain. Conversely, warm air moving poleward will lose some of its heat on the way, either by contact with colder ground or by radiation into space. Thus, some of the eddy potential energy resulting from the side-by-side location of cold and warm air masses will be lost before it has a chance of being converted into eddy kinetic energy (Figure 60).

On the other hand, as we have sketched in Figure 61, condensation will more often form in warm moist air overriding a wedge of cold air than in the cold air itself. Each gram of water uses approximately 590 calories of energy as it evaporates. (That is why you feel cold when you emerge from a swimming pool, and that is why perspiration helps you to survive desert heat.) The same amount of heat is given off to the surrounding air as water condenses into cloud droplets. As droplets freeze into ice crystals, another seventy or so calories are released per gram of water.

Thus, we see that the processes of condensation will add heat to the atmosphere. Since this heat is hidden within the water vapor before condensation, and we don't sense it as "hot," we call this heat the *latent heat of condensation or sublimation* (the former if water droplets are forming, the latter if ice crystals are generated from vapor). In contrast, *sensible heat* is the heat felt in plain, dry, hot air. If more latent heat is released in the warm eddies than in the cold ones of our eddy potential energy distribution, the warm air will get warmer. Thus, the latent heat release will contribute to the generation of eddy potential energy and hence to eddy kinetic energy.

Storms and Hurricanes

If we take a look at our most powerful eddies, our big cyclonic storms, we find that either the conversion of mean potential energy or the release of latent heat may be important sources of the storms' kinetic energy—that is, their devastatingly high wind speeds. The blizzards of the Great Plains draw their force mainly from the clashing of cold Arctic air out of Canada with warm tropical air from the Gulf of Mexico. The large temperature differences between the two air masses allow for the conversion of large amounts of potential energy into kinetic energy. Invariably, these storms are associated with strong jet streams at tropopause level, which blow mostly from a southwesterly direction and "steer" the storm centers along the same heading from southwest to northeast.

The tropical cyclones, better known as *hurricanes* (*typhoons* in the Pacific) are, on the other hand, an entirely different breed. Temperatures in the tropics do not vary greatly. Therefore, little mean potential energy is avail-

able to feed those forceful eddies. However, the torren-
tial rains associated with these storms, and concentrated
around the center or *eye* of the storm, by release of
latent heat of condensation provide more than enough
energy to keep the cyclone going.

An easterly current at tropopause level, attaining at
times jet stream strength, carries these tropical storms.
Hence the preferred direction of hurricanes and typhoons is from east to west. If such a tropical cyclone
clashes with a cold air mass of polar origin, it also col-
lides with the westerly jet stream associated with the
front between cold and warm air. This jet stream will
force the tropical storm to curve toward the northeast.
Whenever hurricane, or typhoon, flags are flying, people
on the East Coast, or the Japanese, anxiously look for
this veering of the storms.

Once in a while the westerly jet stream intrudes into
low latitudes which normally are "reserved" for the easterly winds aloft that carry the tropical storm. When such
an intrusion occurs, the hurricane may be caught in a
giant vortex, with the westerly jet-stream flow on one
side, and the equatorial easterlies on the other. As long as
these two opposing currents are well matched, the hurricane may find itself in a tug-of-war. It will follow a
seemingly erratic path, playing hide-and-seek with radar
crews and reconnaissance aircraft, and keeping millions
of anxious watchers glued to their television sets, fretting
for the next weather bulletin.

Once the westerly jet has gained its victory—and in
the long run it usually does—and the storm has started
to curve toward the northeast, it may incorporate into
its vortex some of the cold polar air. Now a new reservoir
of potential energy previously not available may be
opened, and the hurricane or typhoon may once more
gain in fury. Since it now carries a system of cold and
warm fronts, resembling a cyclonic storm of temperate

latitudes, it no longer is considered to be a tropical storm.

On the other hand, if a hurricane hits land (say, over the Mexican or Texas Gulf Coast) before it curves, it finds itself cut off from its lifeline, the moisture evaporating from the churned-up ocean surface. This moisture, condensing to heavy rain clouds, provides the vital latent heat of condensation. Thus, the storm is doomed to die, but not until a nightmare of destruction is left along the coastal plains.

More about Energy

Let us return for a moment to Figure 60. The conversion of eddy potential into eddy kinetic energy is the main accomplishment of the polar front jet, since in its domain we have fronts along which warm and cold air masses can easily slide up and down.

These fronts are associated with cyclones, as we have seen on the preceding pages. With cold air sinking to the rear of the cold front, and warm air rising over the warm front, potential energy is released by lowering the common center of gravity of both air masses. This potential energy is made available to the jet stream in the form of kinetic energy. As we have seen before, cyclones are associated with troughs aloft: The jet stream flow is not running smoothly from west to east, but follows a meandering pattern, portraying to a certain extent the shape of the frontal lines on the earth's surface (Figure 71 in the following chapter). Thus, the kinetic energy generated by the juxtaposition of large cyclones and anticyclones will be mainly of the eddy variety.

But what happens now to the eddy kinetic energy? Friction at the ground will consume part of it, and part may feed back into the eddy potential energy reservoir. We observe this process many times on the weather map. A big trough develops, let us say, over the midwestern United States. It forces the jet stream into a

wide meander, which in turn will disturb the flow over
the Atlantic and over Europe into more meanders.

Part of the eddy kinetic energy will be used to feed
the mean kinetic energy, and will be consumed by friction there. As was pointed out in Chapter V, the atmospheric circulation oscillates between low-index and
high-index periods. During the former we find large amplitude troughs and ridges in the jet stream flow; during
the latter the jet streams run nearly parallel to latitude
circles. Apparently there are periods in which eddy kinetic energy contributes toward mean kinetic energy
(transition from low to high index), and others in which
zonal motion breaks down into large eddies, indicating
that mean kinetic energy is transformed into eddy kinetic energy (transition from high to low index). The
duration of the periods of transition may be of the order
of two weeks. We can look here for one explanation of
why weather is as fickle as it appears to be.

In the discussion of tropical storms we mentioned the
presence of easterly winds in the tropics near tropopause
level. Isn't their presence in violation of our reasoning
about the northward transport of angular momentum and
the westerly winds ensuing from it? If the circulation in
the tropics and subtropics were truly a simple wheel
symmetric about the earth's axis (Figure 51), carrying
all the required heat and momentum only by meridional
motions, we should expect only westerly wind components aloft. The air motion would resemble a huge spiral:
the rising near the equator; the northward flow aloft
with increasing westerly wind components due to conservation of angular momentum; the sinking underneath
the subtropical jet stream; the return flow toward the
equator near the ground with an easterly (trade) wind
component weakened by surface friction; then again a
rising over the equator; and so on.

In these circumstances, we should expect a uniform

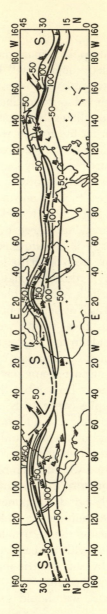

Fig. 62. The subtropical jet stream, shown by isotachs on the 200 *mb* surface of 25 February 1956 is characterized by hemispheric three-wave pattern. (After T. N. Krishnamurti)

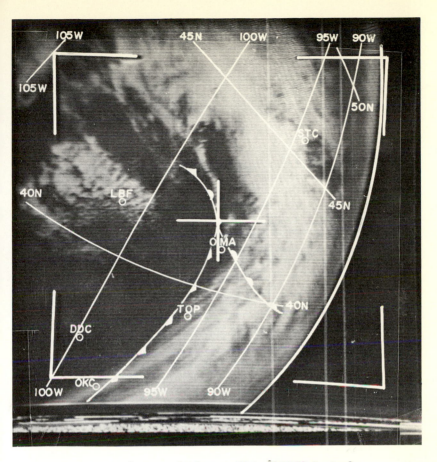

PLATE VII. Tiros V photograph taken on 18 April 1963. Latitude-longitude grid, fronts and positions of radiosonde stations are superimposed. Photo courtesy of National Experimental Satellite Center, Suitland, Maryland.

PLATE VIII. Rotating "dish-pan" generates symmetric three-wave pattern made visible by aluminum particles scattered on water surface. Photo courtesy of Dr. Dave Fultz, University of Chicago.

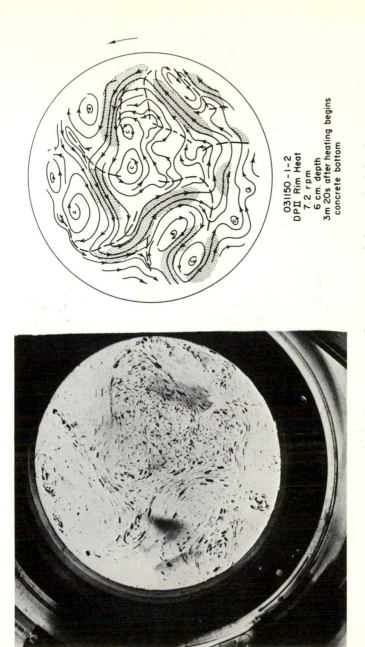

031150 - 1 - 2
DPII Rim Heat
7.2 rpm
6 cm. depth
3m 20s after heating begins
concrete bottom

PLATE IX. Asymmetric tilting waves appear in flow in rotating dishpan, which for this experiment had no cooling cylinder at the center. In diagram shading indicates jet streams, tilt is at left and corresponding streamline pattern at right. Photo courtesy of Dr. Dave Fultz, University of Chicago.

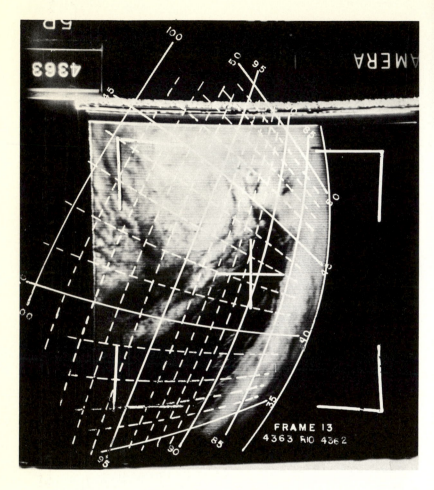

PLATE X. Tiros V photograph taken on 19 April 1963. Latitude-longitude grid is superimposed. Courtesy of National Environmental Satellite Center, Suitland, Maryland.

PLATE XI. Clear air turbulence (CAT) tore off the vertical stabilizer of this B-52 jet bomber in flight. Photo courtesy of The Boeing Company.

PLATE XII. Severe damage to tail of B-52 attests the dangers from CAT but did not prevent pilot from landing safely. Photo courtesy of The Boeing Company.

ring of high pressure to lie underneath the subtropical
jet stream. This is not the case in the real atmosphere.
For one reason, the earth rotates too fast and the atmosphere requires more heat and momentum transport
than such a simple circulation is able to handle. For another reason, large mountain ranges—the Rocky Mountains, the Himalayas, and Andes—force the zonal atmospheric motions into deviating meanders.

The effect of these influences is to break up the high-
pressure belt underneath the subtropical jet stream into
a cellular pattern. Meteorologists talk about a Bermuda
high, an Azores high, and so forth. The fact is that the
subtropical jet stream, especially in winter, describes a
well-developed pattern of three waves around the Northern Hemisphere (Figure 62). Similar conditions presum-
ably prevail in the Southern Hemisphere, although they
are less well explored there for lack of aerological data.

How does this breakup of the meridional circulation
and its associated jet stream affect the upper wind regime? As we see in Figure 63, the northerly flow on the

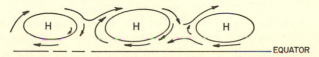

Fig. 63. Subtropical high-pressure belt breaks into several
high-pressure cells.

eastern edge of each individual high-pressure cell will
acquire an easterly component of motion, prescribed by
the tendency to conserve angular momentum and also
by the forces of pressure gradients trying to move the air
along geostropically. Thus, we find east winds in the upper troposphere of the tropics as well, and not only in the
low-level trade winds. Westerly winds prevail in temperate latitudes. The breakup of the meridional circula-

tion into cells actually helps to transport westerly momentum poleward, because mathematically (and physically) a southward transport of easterly momentum is equivalent to a northward transport of westerly momentum, as we have demonstrated in Chapter V.

The Tropical Easterly Jet Stream

In summer these subtropical high-pressure cells are found farther away from the equator than in winter. These pressure cells appear to be locked into a very stable position over the Plateau of Tibet and over the Sahara Desert. Since the air flowing around the eastern edges of these high-pressure areas has quite a distance to move before it reaches the equator, it attains high velocities and blows almost straight from the east. This movement is called the *tropical easterly jet stream*. It reaches its maximum velocity at a height of approximately 16 *km* (near a pressure level of 100 *mb*). From what has been said earlier, you may remember that this is the height at which the tropopause is usually found in the tropical regions. Thus, the tropical easterly jet also qualifies as a tropopause jet stream.

Only the Sahara high and the Tibet high are of sufficient stability in summer to generate a steady and strong easterly jet stream. In other regions of the subtropics of the Northern and Southern Hemispheres (and in other seasons) the high-pressure cells have a tendency to shift position from day to day. Therefore, easterly jet stream velocities are encountered only infrequently elsewhere than over India and Africa in summer. There the steady tropical jet has a tremendous influence on weather. It is intimately linked to the occurrence of the Indian and African *summer monsoons* and the rainy seasons in those parts of the globe. More will be said in Chapter VII about those phenomena.

Since the existence of the tropical easterly jet stream

is tied to breaks in the subtropical high-pressure belt, and to resulting eddy motions, we may argue that the cause lies in eddy transport processes, and not in mean meridional transports, because the latter, directed poleward near the tropopause, should result in westerly winds.

Jet Streams above Tropopause Level

So far we have dealt with jet streams that are relatively easy to explore with our present radiosonde network. Still, they conceal many mysteries though researchers of many countries are slowly but surely unveiling them. As we go higher up in the atmosphere observational data become more and more sparse. U-2 aircraft may bring back wind and temperature measurements from levels as high as 20 *km,* special soundings balloons may reach heights of 40 *km,* and meteorological rockets probe atmospheric motions up to 70 *km* or so. These measurements, however, are costly and therefore difficult to maintain on a routine basis.

Nevertheless, by patient research, scientists have been able to piece together a fascinating jigsaw puzzle of motions in the high atmosphere. There are many pieces missing, but with our hands and minds stretched out into interplanetary space, the forbidding realms of the upper atmosphere become less and less menacing. With astronauts now traversing these regions, we have to learn more about winds and weather in rarefied air to assure their safety.

In describing Figure 56 we commented that the lower stratospheric layers above the tropopause are warm over the pole and cold over the equator. In summer this reversal of the meridional temperature gradient is found throughout the stratosphere. It is caused by the continuous sunshine over the pole. For the winter season we will have to qualify our statement. Near the polar front jet stream at tropopause level we do find a reversal of the temperature gradient—warm stratospheric air poleward of the jet stream, cold air equatorward. As we have pointed out, it is this reversal of the temperature gradient which determines the height at which maximum wind speeds occur. The reversal is not nearly as drastic in winter as in summer. It is caused mainly by upward and downward motions around the jet stream. Sinking motion in the stratosphere to the north of the jet causes the air to warm; rising motion to the south of the jet cools the air.

As we go higher and away from the level of the polar front jet stream near the tropopause, these effects of vertical motion induced by the jet stream itself become smaller and smaller. Radiation effects start to dominate. Now we are aware of the fact that in winter the high latitudes do not receive any solar radiation at all. The higher layers of the stratosphere, therefore, will be much colder over the pole than over the equator. Hence, in the winter hemisphere the meridional temperature gradient will continue to be directed from equator to pole. The inclination of constant pressure surfaces will become steeper and steeper, and the westerly wind velocity will increase with height. It reaches a maximum near 50 *km*, where (Figure 2) the *stratopause* indicates a sharp kink in the vertical temperature profiles.

It has been found that these strong winds also occur in a jet-stream-like band circling the pole. We call it the *stratospheric polar night jet stream* because it occurs only in winter when the sun is below the horizon. In analogy to the tropopause jet streams, we find the polar night jet also to be a pause phenomenon because of its association with the *stratopause* (Figure 2).

In summer the stratosphere is warmer at the same height over polar than over the equatorial regions. We have east winds prevailing at and below the stratopause. These winds usually are too weak and not well enough organized to qualify as jet streams, at least in the stratosphere. In the mesosphere, above 50 km (Figure 2), they may attain considerably higher speeds, but data from this region are still too sparse to allow the mapping of this easterly jet stream in its daily behavior.

Since it is winter in one hemisphere while the other enjoys summer, we find the strange phenomenon of westerly gales blowing at great heights and high latitudes around one pole, while easterlies prevail in the high stratosphere of the other hemisphere. With the shift of seasons this circulation pattern switches from one hemisphere to the other.

Strange Winds over the Equator

Besides the seasonal change in flow direction, the stratosphere has still other surprises in store. When the volcano Krakatoa (in Indonesia) exploded in August of 1883, it sent its ashes up to heights well above 20 km. At these levels, quite unexpectedly, the cloud of ashes circled the globe from an easterly direction at speeds of about 60 knots indicating that there was a steady easterly jet stream blowing at those levels. In commemoration of the discovery, these winds were called the Krakatoa easterlies, long before aircraft and balloons encountered the jet streams at temperate latitudes and lower heights. Only slowly did the ashes spread into higher latitudes, giving rise to beautiful red sunsets.

Easterlies in the stratosphere over the equator should not be difficult to explain. We have seen that the stratospheric circulation changes its direction between summer and winter, requiring a large amount of angular momentum to be transported from one hemisphere to the other. This momentum transport will have to be associated, at least to a certain extent, with mass flow across the equator. If angular momentum tends to be conserved in this flow, easterly velocities should increase as the air approaches the equator; it should decrease again as the air recedes from the equator escaping into the other hemisphere. We might, therefore, expect an easterly wind maximum to occur right over the equator at levels where the cross-equatorial transport reaches its maximum and has its widest latitudinal extent. Early measurements showed the east wind maximum near a level of 35 *km*.

So far, so good—until A. Berson (1859–1942), German meteorologist, made an expedition, in 1908, to Lake Victoria in Africa. He discovered a layer of west winds right over the equator at heights of about 20 *km* or slightly higher (near the 50 *mb* pressure level), with east winds above and below. At only about 20 knots or so, these winds do not attain jet stream velocities. They are puzzling, nevertheless, for how can any portion of the atmosphere rotate faster than the earth's equator? Conservation of angular momentum obviously cannot explain the existence of these winds. They must be driven by pressure forces which, in turn, are caused by a peculiar and rather steady temperature distribution.

While scientists were still puzzling over Berson westerlies, as they are called, a new discovery was made in the early sixties. The layers of the equatorial stratosphere that showed east winds one year, experienced west winds the following year, and vice versa. As this strange cycle kept repeating itself, it was found that the exact period with which easterlies or westerlies reappeared at the same level was actually not two years, but twenty-six months. Furthermore, the belt of east winds or west winds—depending on the year in which you start your investigation—appears first at very great heights and gradually works itself downward, finally losing itself near the tropical tropopause (Figure 64). Puzzles upon puz-

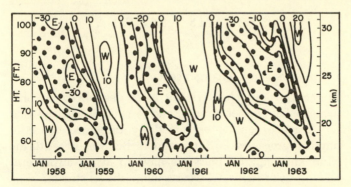

Fig. 64. Alternation of east winds and west winds over Canton Island. (After U. S. Navy Weather Research Facility)

zles! Why twenty-six months and not twenty-four in this mysterious repetition of wind regimes? Why is there a cycle at all? How long will it continue to repeat itself with this off periodicity? For centuries, or just for a couple of sunspot cycles?

Chapter VII

JET STREAMS AND WEATHER

Divergence and Convergence

In the previous chapters we have described the phenomenon of the jet stream, and we have reasoned that its existence can be explained by direct reference to physical processes. So far, so good. But was it worth the effort and paper? Unless we have to fly an airplane at an altitude of 10 *km,* who cares about the jet stream?

The answer is simple: We have not wasted our time explaining the ifs and whys because the jet streams are intimately linked to the weather, and only in understanding *them* can we understand the ever-changing aspects of the atmosphere surrounding us. How do the jet streams do it?

Let us consider air flowing through a typical jet maximum, as shown in Figure 65. From experience, and from

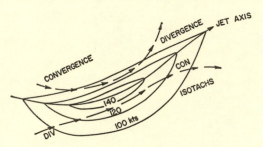

Fig. 65. Convergence and divergence around jet maximum with nearly straight flow.

examples given in Chapter IV, we find that such jet maxima move along at a speed of approximately 10 *mps,* while the air currents themselves have speeds of 50 *mps* and more. Thus, the air actually flows through the speed or isotach pattern, and the latter does not simply swim along with the air. That means that air moving through a jet maximum has to accelerate as it moves in on the rear side; it decelerates as it moves out the front end of the wind speed maximum. The air, therefore, does not flow at uniform speed in the jet stream region. In simple terms we might conclude that where air starts to accelerate, the faster flow will cause a divergence of mass. In deceleration there will be convergence (Figure 66). We

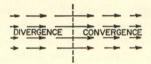

Fig. 66. Divergence and convergence (from left to right) in straight accelerating and decelerating flow.

may compare the effect to a column of marching soldiers: If the first row starts to run, a gap will show between rows. If the first row suddenly stops walking, the rest of the men will bump into it.

Near the jet stream things are not as simple as in Figure 66 because we have the curvature of flow to consider as well. Where streamlines come together in flow of uniform speed, there will be convergence. Where they separate, there will be divergence (Figure 67).

The effects of acceleration or deceleration and of streamlines approaching each other or separating will be present simultaneously in the jet stream region, as indicated in Figure 65. We will have to add both effects to arrive at the true distribution of divergence and convergence about a jet maximum. That is what we find in nature: In the left front quadrant of a wind speed maximum the effect of streamline spreading overwhelms the

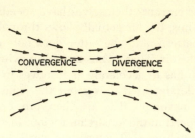

Fig. 67. Convergence and divergence in flow at uniform speed but with streamlines first merging and then fanning out.

effect of deceleration. Thus, we have mass divergence in this area. In the right front quadrant the air by deceleration piles up faster than streamline spreading could help to carry it away. Therefore, we find mass convergence here (Figure 65).

The distribution of flow to the rear of a jet maximum leads, in a similar way, to convergence in the left, and divergence in the right quadrant. If a jet maximum is at the tip of a strongly curving trough (Figure 68) ob-

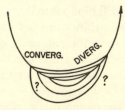

Fig. 68. Around jet maximum in a strongly curving trough of low pressure curvature of streamlines will have overpowering effect.

viously the effects of curvature of streamlines will be even more overpowering in those quadrants which already have shown their predominance in the earlier example of nearly straight flow. The divergence in the left front, and the convergence in the left rear of the wind speed maximum will therefore be excessively well developed. On the anticyclonic right side of the jet axis (where anticyclonic wind shears prevail, Chapter III) the effects of streamline spreading or approaching will offset the deceleration and acceleration effects. Thus, whatever convergence or divergence may be found in this region, it will be very weak.

Let us now consider the effect of such convergence and divergence patterns generated by the jet stream. If, at jet stream level, air is removed from a vertical column by the action of divergence, the total weight of this column, measured from the top of the atmosphere down to the ground, will become less. Since the total weight of the column, however, is equivalent to surface pressure, a decrease in this weight will lead to a fall in surface pressure. Underneath the divergence region caused by the jet stream a surface low will be generated. This low might well be the origin of a storm, if the divergent processes keep alive and if, therefore, the pressure keeps falling for some time (Figure 69).

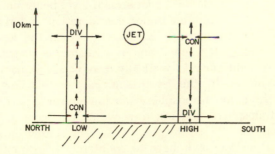

Fig. 69. Schematic vertical cross section normal to jet stream, in exit region, shows distribution of divergence and convergence.

What will happen near the surface? Air will move into the low-pressure region, trying to fill it up. The earth's

rotation will deflect this motion toward the right, trying
to make it geostrophic or parallel to the isobars or contour lines. Because near the ground there is always friction, geostrophic balance is never reached, and instead
of circling around the low, the air will spiral into it.
Clearly, this establishes *convergent* flow in the low-
pressure region at the ground (Figure 70).

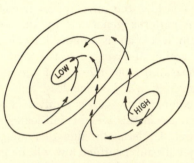

Fig. 70. Under frictional forces acting near surface of earth
air flows out of high-pressure region into low-pressure region.

Now we find the air at jet stream level being pumped
out of the air column under consideration and replenished near the ground by convergence in the low. In
order to avoid the generation of a vacuum at tropopause
level, air will move in vertically to replace the mass that
was removed by the jet stream. Part of this flow will
come from the troposphere moving upward, part from
the stratosphere, moving downward. We have to keep
in mind that the stratosphere is layered stably; thus any
vertical motions will have to counteract buoyant forces
—much more so than in the troposphere. Most of the re-
plenishing air masses, therefore, will come from the
troposphere (Figure 69).

Conditions on the other side of the jet stream are analogous. Here we find convergence of flow aloft; air is

piled into a vertical column, increasing the weight of the total column and hence increasing the surface pressure. Air will flow out from the ensuing high-pressure area, curving anticyclonically because of the deflecting force of the earth's rotation.

Looking at Figure 69 we now see a continuous pump that transports air up and down in the atmosphere. We should not forget, however, that while these vertical motions are in progress, the air moves horizontally in the jet stream at very high speeds. Thus, air that rises in the region of the surface low actually does not end up in the high-pressure region south of the jet, but rather far downstream. Besides, we should keep in mind that the divergence-convergence distribution shown in Figure 69 holds only for the front end of a jet maximum, the so-called *exit region,* because air exits from the maximum. (We assume the jet in this diagram to blow into the plane of the cross section.) The rear end of the jet maximum, or its *entrance region,* carries a distribution of DIV and CON opposite to the one shown in this figure.

With sinking motion underneath convergence at jet stream level, and rising underneath divergence, and with the strong speeds prevailing near the wind maximum, we have the situation shown in Figure 71.

The air originating from the left rear quadrant of the jet maximum sinks underneath the jet maximum and ends up in the high-pressure region of the right forward quadrant (curving anticyclonically in this region). The air from the right rear quadrant, on the other hand, rises over the jet axis and moves into the left front quadrant, curving cyclonically.

What we have said has far-reaching implications. We are now in a position to describe and understand what happens when a cyclone forms. As a matter of fact, with our present knowledge we should be able to develop a

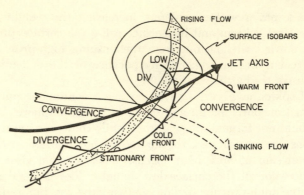

Fig. 71. Jet axis and surface fronts are shown schematically here with the divergence and convergence regions at the jet stream level. Arrows indicate rising and sinking flow.

quite sophisticated cyclone theory. First of all, we come to recognize that no cyclone will form along a mid- or high-latitude frontal zone *without* the presence of a powerful jet stream aloft. What else could provide the essential divergence of flow aloft? The divergence, remember, acts as a pump providing the necessary fall in surface pressure. Once such an area of pressure fall is well organized near an existing surface front, the air motions sketched in Figure 70 set in. Spiraling flow, characteristic of a cyclonic storm, is taking its course.

In the years following World War I the Norwegian meteorologists had a fairly good understanding of the action in the big storms sweeping across the Atlantic. Vilhelm Bjerknes, a Nestor of modern meteorology, considered that a clashing of two "enemy" air masses—one warm, the other cold—along a well-defined boundary line was the main essence of storm development. When he promulgated his cyclone theory, he obviously had fresh in his mind the battlefields of northern France, where armies bled to death in vicious combat along the "front" line. The analogy with air mass "fronts" is quite striking.

Is there not a continuous battle raging between cold and warm air? Rushing advances of a cold outbreak overrun the "entrenched" positions of warm air, while along a different front line warm air pushes its victory far to the north. Instead of the dreaded artillery barrages of World War I, we have rain squalls, high gusts and thunderstorms, maybe even tornadoes, as the bombardments of the fighting air masses.

The Norwegian cyclone theory did not yet take into account the pumping effect of jet streams aloft. Maybe World War II helped to shape our thinking again, even in meteorology. If World War I gave us the concept of a "front," World War II offered the idea of "dynamic forces" pushing rapidly ahead in "blitz" fashion, encircling enemy forces and obliterating them in their cut off positions. Our thinking about eddies in the jet-stream flow, about vortices cut off from the main westerly circulation and slowly losing their identity as cold or warm pockets of air, certainly bears some resemblance to what may have gone on in the minds of generals and field marshals traveling the dusty roads of Europe and North Africa.

We should not be too skeptical about this analogy between modern warfare and meteorology. Since scientists are human beings, why should their minds and their ways of thinking not be molded by such traumatic experiences in the course of human tragedy?

Aside from the frontal concept, the Norwegian view of cyclone development produced another concept which proved extremely valuable to modern meteorology. In their description of the life cycle of a storm they usually started with an undisturbed state in which cold and warm air masses "lived peacefully" side-by-side, separated by a straight front line. Each air mass flowed along on its proper side of the front, the flow paralleling the extent of the front line. Suddenly, out of nowhere, a small disturbance causes the warm air to produce a small "dent" in the straight front line: A so-called *wave disturbance* is born. It starts to spread and amplify, especially as surface pressures begin to fall near the apex of the newborn wave, and as air motions try to spiral into the low-pressure region. The small *frontal wave* shown in the lower left corner of Figure 71 closely resembles such a growing disturbance.

As this wave develops, the originally stationary front now starts to advance. Along one portion of it cold air tries to push ahead, marking the so-called *cold front* (represented by triangles on the weather map). Along the frontal portion to the east of the wave apex warm air starts to advance, marking the so-called *warm front* (semicircles on the weather map).

Since the cold front usually moves slightly faster than the warm front, it tends to catch up with the latter. Thus, as the storm grows deeper, the cold front becomes more pronounced. The mature storm shown in the upper right of Figure 71 may serve as a typical example. The small portion of the cold front extending beyond the apex of the wave actually indicates how far the cold front has overrun the warm air contained in the so-called *warm sector*. This process of cold air overrunning the warm sector is called *occlusion*. As the occlusion proceeds, the warm air is lifted off the ground, and cold air slides in underneath. Therefore, the reservoir of potential energy, which requires warm air to lie side-by-side with cold air, is exhausted in the occlusion process. The storm is doomed to die.

It was the experience of the Norwegian meteorologists who watched frontal zones and cyclones moving across the Atlantic and the North Sea that a number of wave disturbances usually traveled in sequence along one well-established frontal zone. As the first wave starts to mature into a full-grown cyclone, a second wave dis-

turbance appears, as indicated in Figure 71. In the Norwegian terminology the original cyclone has spawned a
"daughter cyclone." From our previous reasoning we now
understand that the daughter cyclone develops in the
right rear quadrant of the the same jet maximum that
carries the original cyclone. With the deepening of this
second cyclone a new jet maximum will develop which
now can breed a third wave disturbance, and so on.
Finally we are faced with a whole *cyclone family* moving
across the Atlantic Ocean. Looking at one particular
weather map, we may see as many as four or five such
waves strung out along a frontal zone. Each wave shows
more advanced stages of development as we follow the
extent of the frontal zone eastward.

Over the United States we usually cannot observe the
development of cyclone families. The huge barrier of the
Rocky Mountains disturbs the jet stream flow so much
that one large storm or blizzard, and possibly one additional small wave in its wake (Figure 71), consumes
most of the available potential energy of the cold outbreak sweeping down from Canada. There is usually
nothing left for a family of cyclones to grow on. We are
no better off, however, than the Norwegians. These blizzards and their associated cold outbreaks are bad
enough, even though they don't beget a whole family.
The fruit growers as far south as Florida have an expensive story to tell on this subject.

The problem that arose from the concept of a wave
disturbance which amplifies into a full-blown cyclonic
vortex was to find unstable flow processes in the atmosphere that could lose their balance from an ever-so-slight
push or, in scientific terms, by an infinitesimal disturbance. As it turned out, the strong wind shears which we
find underneath a jet stream within the frontal zone provide such a setup. A mountain range over which the jet
stream is forced to run may disturb the flow, and the

disturbances show a tendency to produce a deep dent in the originally straight frontal zone and its associated jet stream. The dent grows larger and larger, seemingly all by itself, until we finally have a fully grown cyclone at hand. We just mentioned that the corrugated terrain of a mountain landscape may offer the initial disturbances which cause all this havoc. Small wonder that we find so many deep cyclones and blizzards right downstream from (to the east of) the Rocky Mountains.

This idea of an amplifying wave disturbance has been greatly exploited. The mathematical equations governing atmospheric flow were fed into huge electronic computers. By solving these equations numerically, meteorologists could compute forecasts of the atmospheric flow patterns. When small mathematical disturbances were fed into the computer, the stability of the flow patterns could be tested: If nothing much happened to the forecast solutions, the atmospheric flow evidently was stable. If, however, the forecast flow patterns all of a sudden started to bend and twist into big troughs and ridges, the original flow conditions were unstable, and liable to break down under the slightest push.

Numerical modeling of the atmosphere, as we call this art of describing atmospheric behavior by equations which a computer is willing to accept, has come a long way. We are able to generate cyclones and anticyclones out of originally straight flow. All this without spilling any water into or from a dishpan, without even looking at a weather map. All it needs is a large, super-fast electronic brain with a high-speed printer hooked to it. Frontal zones are an entirely logical concept, which such a computer is capable of "thinking up" from its equations. And so are jet streams. This means that we can actually "generate" jet streams, not only in the model experiments described earlier, but also on the print-out paper of a computer.

This sounds like a fascinating step toward a full understanding of the atmosphere. Let us be a little cautious, though. True, by generating realistic jet streams out of nothing but equations, we at least know that these equations are important in describing the atmosphere. Equations, however, do nothing but weigh the size of one term against the size of another one. They don't tell us which comes first, the chicken or the egg. And this is precisely our biggest problem which even computers haven't answered yet. How can we say with certainty which was the cause, and which the effect? To appreciate our dilemma, think about this:

Does the jet stream make a cyclone? Yes! Because the jet's divergence and convergence pattern provides the necessary pumping mechanism, as we have seen in Figures 65, 69, and 71. Does a cyclone make a jet stream? Yes! Because the potential energy released in a cyclone through sinking of cold air and rising of warm air, especially during the occlusion process, provides the necessary source for the kinetic energy of the jet stream. Now, which is the chicken and which the egg: the cyclone or the jet stream? Don't worry. The latest computer has not found the answer yet either.

In spite of this dilemma, our research allows one conclusion which is quite relevant: *There is no cyclone without a jet stream; there are, however, jet streams without cyclones.*

Motions near the Polar Front

We have gone to great length to show that polar front jet streams in mid-latitudes are associated with frontal discontinuities between air masses. If pressure falls occur along such a front, usually a cyclone develops. This is indicated in Figure 71, where a storm is located underneath the divergent area of the left front quadrant, and a smaller wave disturbance forms underneath the divergence of the right rear quadrant.

Upward motion in the atmosphere is associated with decreasing pressure along the air trajectories, hence with expansion of the air and cooling. Frequently this cooling is strong enough to produce condensation of water vapor and clouds. We should expect, therefore, that clouds will be found along the ascending branch of motion in Figure 71. The descending branch should be cloud-free, because during the compressional warming any cloud droplets would tend to evaporate.

Weather satellites, such as the TIROS, NIMBUS, and ESSA series, present an excellent opportunity to check on these expectations. Plate X shows a photograph taken by TIROS V and Figure 72 gives the associated wind speed analysis at jet stream level. The distorted outline of the area covered by the photograph is superimposed upon this analysis.

By comparing the two figures we can very well identify the cloudy band associated with upward motion. The clear band with downward motion dips underneath the moist band, according to Figure 71. From there on, the band can no longer be identified, because the clear air is hidden underneath a cloud band. An aircraft flying through that region might, however, encounter a cirrus overcast near tropopause level, and a clear space below.

In the region of the big storm we see a large area of dense clouds in the TIROS photograph. These clouds occur in the lower levels of the atmosphere, and are produced by the upward motion shown in Figure 69. Low cloud layer, dry area of sinking motion, and high-level cloud band produced by rising motions near the jet will all contribute to the peculiar spiral appearance of a typical storm as seen in satellite photographs.

With the advances made in satellite technology we not only are able to "see" cloud patterns from spacecraft alti-

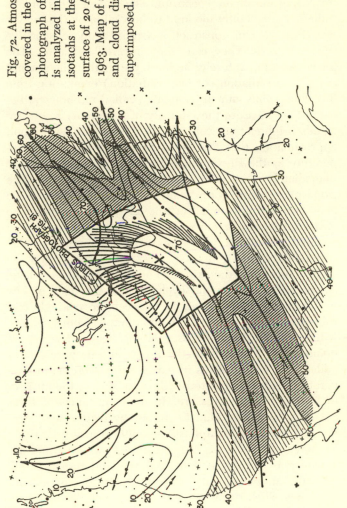

Fig. 72. Atmosphere covered in the TIROS photograph of Plate X is analyzed in chart of isotachs at the 250 mb surface of 20 April 1963. Map of area and cloud distribution superimposed.

tudes, but also can measure the approximate heights of their tops. Water, especially in droplet or crystal form as contained in liquid or ice clouds, radiates in the infrared portion of the spectrum. The wavelengths of this radiation are slightly too long to be visible by the human eye. But we can construct instruments more sensitive than our eyes. They can "see" infrared radiation. The lower the temperature of a cloud, the smaller the amount of infrared radiation (or heat radiation) emanating from the cloud. Thus, our instruments (called *radiometers* because they measure the intensity of radiation) see warm clouds brighter than cold clouds. If we assume in good approximation that the cloud tops have the same temperature as the surrounding clear air, we may conclude immediately that cold clouds reach higher into the troposphere than warm clouds. The exact height we may obtain by comparing the cloud-top temperature measured by the satellite (obtained from the measured infrared radiation) with the vertical temperature distribution obtained from the nearest radiosonde station.

As it turns out, the existence of the band of high-level clouds spiraling into the cyclone, and the deck of low clouds, with cloud-free air in between, is strikingly corroborated by the infrared-eye photographs from satellites.

In condensation processes latent heat, as we have pointed out, is added to the ascending air parcel. This added heat compensates in part for the decrease in temperature that is associated with expansion as the air rises to levels of lower pressure. In many instances it turns out that the heat added to the air by water vapor condensing into cloud droplets is enough to make the rising air parcel *warmer* than the surrounding air. Warmer air, however, is less dense than cool air. Under such conditions the air parcel will continue to rise out of its own buoyancy. Such rising air bubbles are magnificently visible in the cauliflower shape of cumulus clouds (Plate

I). If the buoyant energy of the rising bubbles of moist and cloudy air is large, these bubbles may ascend to great heights, giving rise to thunderstorms.

As we have seen in Figure 71, the jet stream region is a place where rising motions and cloud formations are found on a grand scale. The statement made on page 12, that thunderstorms frequently form near the jet stream, finds therein its logical explanation.

If we were to cut a cross section normal to the flow near the exit region of the jet stream, we now could supplement the simple view presented in Figure 69 by the more sophisticated pattern given in Figure 73. We have rising motions and low clouds near the low center. We find a frontal zone separating cold from warm air. Peculiarly enough, the upper portion of this frontal zone does *not* represent a mixture between the two air masses, but contains the dry, cloud-free, and sinking air, shown (Figure 71) crossing underneath the jet axis. This air has actually come out of the stratosphere in the left rear of the jet maximum. We call this upper portion of the front the *jet stream front* because it is linked intimately to the motion processes around the jet stream.

Above the front we have rising processes in conformity to the rising branch shown in Figure 71. Clouds, precipitation, squall lines, and thunderstorms may be associated with this upward motion. Farther south from the jet stream, at last, we find fair weather associated with the high-pressure region.

The stratospheric air intruding in the jet stream front was found to be high in radioactivity in the years of atmospheric nuclear testing. Originally it was thought that radioactive debris suspended in the stratosphere would stay there for a long time because of the great stability and the lack of heat convection, which in the troposphere accounts for most of the vertical mixing processes. As it turned out, however, the jet stream is able

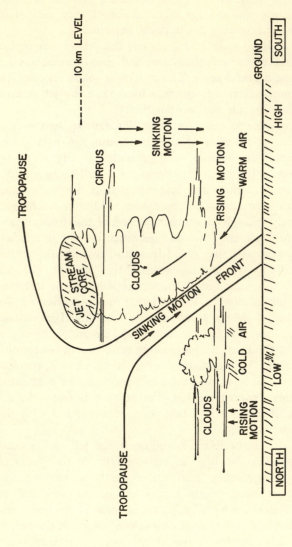

Fig. 73. Polar front jet stream in cross section is pictured schematically to show distribution of vertical motions and clouds.

to remove large quantities of stratospheric air, and its associated radioactive debris, into the troposphere via the jet stream front. From here on the debris reaches the ground by washout in rain or snow, or by settling in the anticyclone equatorward from the jet stream.

Unfortunately, the importance of the jet stream in carrying atomic debris was discovered only after the first damage had been done. Several cases of relatively strong radioactive fallout—coming seemingly out of nowhere— are on record. (Fortunately, even this fallout remained below the level of contamination doctors presently consider dangerous.) The zeal with which nations on either side of the Iron and Bamboo Curtains started to toy with their nuclear firecrackers prompted the Atomic Energy Commission to set up a large monitoring network of observation stations. At each station air is sucked through fine-pored filter paper. A Geiger counter measures whatever radioactivity has been collected.

There was the fallout of September 17 to 20, 1961, when filter samples collected from the eastern seaboard and the Gulf states sent the Geiger counters clicking up to 800 picocuries* per cubic meter of sampled air. (Readings of one picocurie or less were considered normal, even with all the nuclear testing that had gone on during the past years.) Where had this dramatic increase in radioactivity come from? Was it one of the underground tests in Nevada? If so, how could the radioactivity have escaped in such large proportions? Unexpectedly, there came another clue. At about the same time, several American scientists were launching large balloons from a Canadian station near The Pas. These balloons were equipped to measure the intensity of cosmic radiation.

---

* One curie is that quantity of radioactive material that undergoes 3.7×10^{10} disintegrations per second. One picocurie is 10^{-12}, or $\dfrac{1}{1,000,000,000,000}$ of one curie.

Two flights went completely haywire on September 14 and 15. Instead of measuring the nuclear radiation associated with cosmic rays and expected at high stratospheric levels, the recorder traces obtained from the balloons showed unbelievably strong peaks of some mysterious radiation at levels less than $3\frac{1}{2}$ kilometers. What had gone wrong?

Nothing. By sheer coincidence the American balloons had intercepted an invisible cloud of radioactive debris which had traveled from a Soviet test conducted on September 10 over Novaya Zemlya. The debris cloud rode the jet stream until it reached the Arctic coast of Canada. Then it started to descend through the stable layer underneath the jet stream. It had just made it down to about $3\frac{1}{2}$ kilometers when the cosmic-ray balloons intercepted it. Three days later the first debris particles started hitting the ground in the New York region, and from there on the nuclear contaminant continued to settle over the southeastern United States.

There is another interesting case on record. On May 13, 1962 all of a sudden the milk from the Wichita and St. Louis milksheds went "hot," meaning that excessive amounts of over 600 picocuries per liter were picked up in dairy samples taken from these milksheds. (Since strontium 90, a radioactive product of fission bombs, is taken up by human bones through milk, a continuous watch is kept on radioactivity levels in dairy products.) As it turned out, nine days earlier, on May 4, a nuclear device of American manufacture was exploded near Christmas Island, about two degrees north of the equator, in the middle of the Pacific Ocean. Under normal conditions the nuclear debris from such an experiment would travel a long time before it reached the United States. In the present case, however, the jet stream did not behave according to the textbook. Westerly winds in the upper troposphere were present over Christmas

Island on May 4. These winds swept the debris into a southwesterly jet stream that crossed the Pacific at a rapid pace. Sampling aircraft found a "hot" cloud of nuclear debris on May 8 over the Rocky Mountain region at a stratospheric height of 50,000 feet. Normally, nuclear debris traveling at such levels would not bother anybody at the ground. In the present case, however, the upper flow pattern triggered off severe thunderstorms in the Wichita and Kansas City regions the evening of May 8 and the early morning of May 9. The cauliflower tops of these storms actually penetrated the tropopause, and radar crews reported radar echos from as high as 57,000 feet. (This means that cloud droplets or snow crystals large enough to reflect radar waves were present at these levels.) Thus, the thunderstorms penetrated into the hot debris cloud, and the water and ice contained in the cumulus clouds exercised a cleansing effect by washing out the debris from the atmosphere.

From there on the story is simple. Down poured the rain and hail—radioactive from its contact with the debris in the stratosphere. It splashed over meadows and pastureland. The cattle ate the grass, and with it the radioactive debris. The debris (mainly iodine 131) went into the milk, and thence to the dairy plant. There the whole thing was duly pasteurized, homogenized, fortified, modified beyond recognition, and labeled "Grade A." It was delivered right to your doorstep—iodine and all. Luckily, radioactive iodine 131 has a half-life of only eight days, meaning that after this time only half the original amount of radioactivity will be present. Thus, if you did not get your milk fresh from the dairy farm, chances were that most of the radioactivity had decayed by the time you opened the bottle.

Fortunately, such alarming stories have become infrequent since the United States and Soviet Russia decided that the atmosphere was something to breathe and not to contaminate. Ever since the Test Ban Treaty the nuclear contamination of stratospheric air has been on a rapid decline. Unfortunately, however, common sense and responsibility toward unborn generations are not universal attitudes. So some atmospheric testing of nuclear devices is still carried on by other nations.

It is not man-made nuclear debris alone that is transported by the jet stream. We mentioned in Chapter I that high-energy ultraviolet radiation produces ozone in the high regions of the stratosphere. Some of this ozone is also carried downward in the jet stream and reaches tropospheric levels by traveling through the stable layer of the jet stream front. It finally is destroyed on contact with the ground.

From all this we have to conclude that the tropopause is not a rigid separation surface between stratosphere and troposphere. Each jet stream provides its own tropopause gap (Figure 73) through which stratospheric air flows out into the troposphere. The reservoir of stratospheric air is replenished again when air flows in the upward-motion branch shown in Figure 71.

Let us, again, look at Figure 69. Under the divergent region in the left front quadrant of the jet maximum we find upward motion, hence cooling of the air; under the convergence region we find downward motion, hence warming. Drawing a frontal zone, as in Figure 73, we would infer that the cold air would get cooler, the warm air warmer, and the temperature contrast between the two air masses steeper. If that were so, the pressure gradient between the high and low area should increase, and consequently, the wind speeds in the jet maximum should pick up. This is precisely the mechanism by which the jet maximum shown in Figure 65 moves slowly eastward. The temperature and pressure gradients around the front steepen through the action of vertical motions produced by the jet maximum itself.

We may look at this problem at a still different angle: If cold air becomes cooler and warm air warmer, the *potential* energy in this region is obviously on the increase. How can it be? At the expense of *kinetic* energy! The air flowing through the jet maximum in Figure 65 moves faster than the isotachs (lines of equal wind speed) and the jet maximum. Thus, the air decelerates (that is, loses kinetic energy) as it shoots out the front end of the jet maximum. This *kinetic energy* is now expended to build up *potential energy* by steepening temperature and pressure gradients across the front, thus helping the jet maximum to proceed slowly downstream.

In the entrance region of the jet maximum the opposite situation is found. Underneath the convergent region (Figure 65) we have sinking, which, according to Figure 71, occurs in cold air. The rising underneath the divergent region occurs in warm air. Sinking of cold and rising of warm air result in loss of potential energy. Truly enough, this potential energy is converted into kinetic energy as the air accelerates when rushing into the jet maximum. At the same time, sinking motion will cause the cold air to warm, and rising will cause the warm air to cool. Thus, the temperature and pressure gradients across the front to the rear of the jet maximum will tend to relax.

We now can see that the frontal zone with which the jet maximum is associated is continually built up, and added to, in the exit region of the wind maximum, and continually eroded and dissipated in the entrance region. This completes the picture of the slow downstream propagation of a jet maximum.

What has been said so far explains most of what is presently known about jet streams. The correlation of jet streams and weather becomes a simple matter of deduction from our foregoing reasoning. Wherever the jet streams induce upward vertical motions in the troposphere we stand a chance of cloudy weather with precipitation, depending on moisture content and stability of the atmosphere. If the former is high and the latter low, we may expect thunderstorms, squalls, and other forms of violent weather.

Underneath convergence produced by the jet stream, downward motions and hence fair weather should prevail. Since this is the average state of affairs underneath the subtropical jet stream, small wonder that there is little precipitation in the subtropical high-pressure belt. Irrigating the Sahara underneath this high would not increase the precipitation there. First, we would have to shift and relocate the subtropical jet stream before natural rains would fertilize the desert. But how? Perhaps by shifting the Rocky Mountains or the Himalayas to a different location? After all, they are known to distort the hemispheric jet stream flow into huge meanders, thereby affecting the distribution of climate as we presently experience it.

The Monsoons

If the tropopause jets of temperate and high latitudes can be held responsible for the weather in these regions, we may look for a comparable agent at the low latitudes. In Chapter VI we ran across the tropical easterly jet stream, which is found near the 100 mb level (or at a height of approximately 16 km). It occurs only in the Northern Hemisphere during summer and blows persistently over Southeast Asia, India, and Africa.

Figure 74 shows a cross section through the atmosphere running from Trivandrum, at the southern tip of the Indian subcontinent, to New Delhi, in northern India. The radiosonde measurements from which this analysis was made were taken on July 25, 1955, in the Indian summer monsoon period. The core of the easterly jet stream is located over Madras.

In place of an analysis of the actually measured temperatures in this cross section, the temperature *deviation* from the *mean* conditions in the tropics have been entered in the form of *isotherms* (that is, lines of equal temperature). This procedure has an advantage. As we have stated before, horizontal temperature differences are small in the tropics. Thus, the temperature distribution in this cross section would be dominated entirely by the normal decrease of temperature with height. From previous reasoning we have seen, however, that it is the *horizontal* temperature contrasts which control the pressure gradients which, in turn, drive the jet streams. Therefore, if we subtract the normal temperature distribution (decrease of temperature with height) from the actually measured temperatures, we are left with small residual anomalies. These anomalies may now be considered as controlling horizontal temperature and pressure gradients and, consequently, also the geostrophic winds.

Looking at Figure 74, we find that the air over Jodhpur is 15° C warmer than average conditions in the tropics; over Trivandrum, on the other hand, it is just barely colder than average. The greatest anomalies are found near the 200 *mb* level (40,000 feet or 13 *km*) below the jet stream core. Thus, we have a temperature distribution with warm air in the north and cold air in the south, close to the equator. Since this distribution prevails throughout the troposphere, we should expect geostrophic east winds to increase with height. This increase is precisely what is observed.

Above the jet stream level we note a reversal of the temperature gradients; an anomaly of +5° C over the equator and less over northern India. Here again, as in mid-latitudes, the jet stream level is controlled by the horizontal temperature distribution which reverses above the tropopause, making the tropical easterly jet stream one of the family of tropopause jet streams.

166

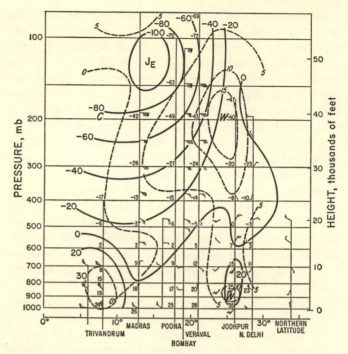

Fig. 74. Tropical easterly jet stream over India on 25 July 1955 is analyzed in cross section. Isotachs (solid lines) are labeled in knots, negative signs indicating easterly winds. Departures (in degrees Centigrade) are shown for isotherms. (After P. Koteswaram)

We also notice in Figure 74 that there is no front underneath this jet stream. Conditions are very similar to those prevailing under the subtropical jet. The difference is that the latter blows from the west and occurs near the height of 12 *km;* the other blows from the east near 16 *km.* With the absence of fronts there will be no frontal cyclone over India. Precipitation will fall mainly in heavy showers, thunderstorms, and squalls.

The seasonal change in weather over the Indian subcontinent has interested scientists for many decades, not only out of curiosity, but because the livelihood of almost 600 million people depends on crops irrigated by monsoon rains. Even the Arabs sailing the Indian Ocean were familiar with the seasonal shift of winds. Their word *mausim*, from which "monsoon" has been derived, actually stands for "season." Only in the late fifties was it discovered how the jet streams shape and influence the monsoonal weather patterns.

The subtropical high-pressure belt, over which the subtropical jet stream moves, undergoes some seasonal migration. In winter we find it at approximately 30° N; in summer it is located close to 40° N. The enormous size of the Eurasian land mass and the large area of high plateau country there influence the migration.

Solar radiation is absorbed and contained within a shallow layer of solid earth which has a low specific heat. In the oceans, which have a high specific heat, radiation penetrates deeper and the heat is mixed down to greater depths. Consequently, land masses will reach higher surface temperatures—and they will attain them quicker—than large water bodies do, and so will the air close to the earth's surface.

From what we learned in Chapter VI, we can understand why the subtropical high-pressure belt is attracted in summer to the Asian land masses where temperatures are high. This effect is accentuated by the heating of the plateau region, which will be warmer than the same levels of the free atmosphere over India (Figure 75).

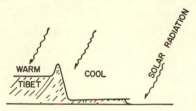

Fig. 75. Solar radiation on Plateau of Tibet makes air warmer than it is in free atmosphere over India.

This situation leaves us with a temperature distribution of warm to the north and cold to the south, calling for east winds aloft over India, as we have found them in the tropical jet streams. The sketch of Figure 75 would lead us to expect the warmest temperatures over the Plateau of Tibet. Actually, however, according to Figure 74, the warmest temperatures are found over northern India. There must be some additional heat source, other than solar radiation, which controls the position and strength of the tropical easterly jet stream.

This source can be identified easily. The southwesterly monsoon flow at low levels, which characterizes the summer season, carries lots of moist air from the Indian Ocean into the continent. So does the flow entering the continent from the Bay of Bengal. Low-pressure disturbances traveling with the easterly flow aloft generate tremendous amounts of precipitation over the Ganges and Brahmaputra Plains. The release of latent heat from the condensation processes leading to the observed precipitation provides the warm temperatures found aloft over northern India.

Thus, we may formulate the following model of the summer monsoon over Asia: With the heating of the continents during spring, the subtropical high-pressure belt shifts northward. The heating of the Plateau of Tibet provides a preferred location for a high-pressure cell to settle down. The tropical easterly jet stream ensuing from this high-pressure cell by southward flow under conservation of angular momentum has its right rear quadrant (with divergent flow near tropopause level) over northern India (Figure 76). The low-pressure region forming under the divergent area aloft provides for inflow of moist air which, when lifting and condensing its water vapor, supplies a large amount of latent heat. This last process is now able to retain the high-pressure region

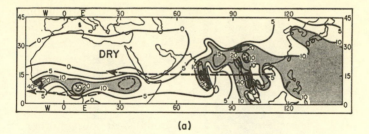

(a)

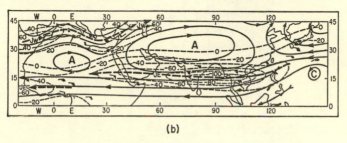

(b)

Fig. 76. UPPER: Isotachs (in knots) and streamlines at 100 *mb* surface 25 July 1955. Heavy dashed lines show jet axes. LOWER: Mean July precipitation (in inches) and positions of jet axes. A's indicate anticyclones; C's, cyclones.

in its position, and to drive the easterly jet stream and the monsoon circulation over India.

During autumn the continents cool more rapidly than the oceans do because the former have stored their heat in a much shallower layer at a lesser heat capacity than the latter. In this process, again, the Plateau of Tibet leads the way. It will suddenly become colder than the free air over India at the same elevation. This contrast calls for west winds to appear over northern India and the Ganges Plains. The subtropical high-pressure belt by that time has migrated southward.

The change in wind regime occurs rather abruptly within a few days, as may be seen from Figure 77. This

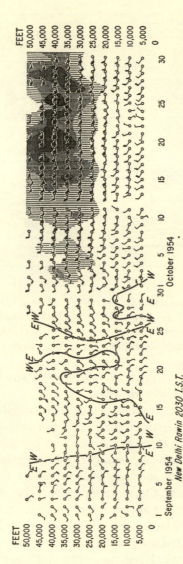

Fig. 77. Time section of winds over New Delhi for September and October 1954. Solid lines indicate division between easterly and westerly components of flow. Different shadings mark wind speeds of 40, 60, 80, and 100 knots.

diagram contains a *time section* of the rawinsonde of
New Delhi for September and October 1954. The vertical
wind distribution on each day has been indicated by
means of little wind arrows. On October 9, for the first
time, westerly winds break through aloft. On October 25
the westerly regime aloft is finally established, and only
a few days later the westerly subtropical jet stream appears in force. It maintains its way until May of the
following year, when it gives way to the upper easterlies
characteristic of the summer monsoon season.

Jet Streams and CAT

So far we have discussed the large-scale implications
of jet streams on weather. There are also lesser atmospheric motions of considerable interest, especially to pilots of high-flying jet aircraft. Every once in a while reports come in that an aircraft has been shaken up badly
in turbulence far away from any clouds. This type of
turbulence, which hits without warning and without visible cause, is called *clear-air turbulence* (CAT). Pilots,
passengers, and airline managers dread it. Pilots may
have a hard time keeping the aircraft under control when
severe turbulence hits. Under prolonged turbulent flight
conditions passengers feel extremely uncomfortable. Several cases are on record where passengers or stewardesses
not tied down by seat belts were bounced against the
ceilings of cabins and injured. Airline managers are not
fond of any kind of turbulence their aircraft might encounter because any stress to which the aircraft structure
is exposed reduces the lifetime of the plane. Wings, sta-
bilizers, and fuselage are subject to vibrations, which may
become quite strong when the aircraft flies through a
turbulent region. The continuous bending of the metal
structure may lead to what is known as "metal fatigue."
(You have probably used metal fatigue many times when

you tried to break a wire without using a wire cutter: You flexed the wire back and forth until it snapped.)

Cases in which aircraft have suffered structural damage from severe CAT are rare, but nevertheless, they are on record. On January 19, 1961 a B-52 crashed near Monticello, Utah. The pilot lost control of the plane in severe CAT. From the way the plane fragments were scattered about the ground it was not quite clear whether the wings had broken off as a consequence of turbulence, or after the pilot had ejected.

On another occasion the vertical stabilizer was ripped off a B-52 when it encountered severe turbulence over the mountains of Colorado. The pilot, not one to panic, flew the plane for quite some time without the stabilizer and managed to land it safely (Plates XI and XII).

It has been known since the late 1950s that CAT occurs more frequently near the jet stream than elsewhere, but not until 1965 were some of the mysterious processes of formation uncovered. When measurements taken over Australia, the United States, and the Soviet Union were pieced together, some of the mystery surrounding CAT was unveiled.

The stable layer of the jet stream front extending underneath the core of the jet stream harbors a good deal of CAT. So does the region in the stratosphere immediately above the jet stream. Stable atmospheric stratification characterizes both regions. Warm air overlies cold air. At the same time strong wind shears are present. Both effects together—stable stratification and (vertical) wind shear—may set off waves, similar to the ones which we observe on a lake when the wind blows over it. Since the density difference between water and air is large, the waves on the lake's surface are very short. The density difference between cold and warm air is very small; consequently, the waves generated by wind shear on the interface between cold and warm air will be long—several hundred meters long, as a fact. Sometimes when there is enough moisture in the air to produce cirrus clouds, we may actually see these waves. They appear as ripples on a sheet of cirrus.

If the interplay between stability and shear is right, these waves may show a tendency to grow, and finally they break up into small irregular eddies. These eddies will be experienced as turbulence and bumpiness when an aircraft flies through this region.

CAT is found frequently over mountainous terrain. The explanation is that the mountain ridges force enough disturbance upon the air flow to set off wave motions which under ideal conditions may reach high up into the stratosphere. These waves usually are in the order of 10 *km* long. If they are associated with clouds (the typical lenticular clouds mentioned with Figure 13 and shown in Plate V), they may even be revealed in satellite photographs.

Such *mountain waves* in the flow aloft usually harbor a large number of small layers with excessive wind shears, which in turn will give rise to CAT in the fashion described. Again, the jet-stream region offers favorable conditions for the formation of such lee waves.

There is still a lot to be learned about CAT. Forecasting techniques, trying to predict its occurrence, are still inadequate. The developers of supersonic transport aircraft to cruise through the stratosphere are paying considerable attention to CAT and its behavior.

Chapter VIII

EPILOG

In the little more than two decades that have passed since man's first encounter with the jet stream we have brought together more knowledge of the atmosphere than had accumulated over all the preceding centuries of human evolution. The science of meteorology reflects the stampede of technology that our generation presently witnesses. The first motor-powered flight of man, the first commercial airliner, the first manned orbital satellite, the first spacecraft on the moon—all have happened within one lifetime.

Similar progress can be mapped in other fields: From radio to television, to microcircuited electronic computers; from Einstein's theory of relativity to nuclear power plants; from Louis Pasteur to the Salk vaccine. Against these whirlwind developments science fiction almost runs the risk of becoming obsolete before the presses can roll.

In those "quiet" days before World War II a scientist could venture a guess, when asked by his friends where a new development might lead. Now he would almost need the imagination of a poet to give an appropriate answer. Before, his answer might have given him the reputation of being a dreamer; now he may be accused of lagging behind reality. Will this pace of progress continue to accelerate? Will it level off? Questions impossible to answer.

What have the jet streams in store for us? Are there still frontiers beyond which we may push into uncharted territory? As an epilog to the facts given in this book, let us risk a glimpse into the crystal ball.

With our present state of technology we are able to chart the course of jet streams near tropopause level, as they meander over the densely populated continents of the Northern Hemisphere. But observing them over the oceans, or over the vast regions of the Southern Hemisphere, is something else. We have no observational network there adequate to bring in data regularly on structure and flow in the upper regions of the atmosphere. To set up such a network would be an unbearable burden on the economy of many of the young nations.

To collect these missing data in an economically feasible way, experiments are presently under way to construct large balloons of tough Mylar, which may stay in the air as long as a year. They will drift with the air currents, and measure temperatures and pressures at flight level. Floating at a constant density surface, they may circle the globe several times before they burst. If they can be tracked by satellites, and their position can be identified from day to day, or from hour to hour, their journey will permit us to calculate wind speeds, and to map the changes in the jet stream patterns around the globe. Still, there is a formidable task left for scientists and engineers. Electronic equipment for these balloons has to be rugged, yet as small as a transistor radio, weighing only a few ounces at the most. Preferably, measuring sensors as well as radio transmitters will have to be printed on the skin of the balloon, thus distributing the solid mass of the equipment over a wide area and minimizing the hazards in a possible collision with an aircraft.

Tracking equipment, either ground-based or in a satellite, will have to be perfected. The cost of the entire balloon and satellite system will have to be cut to reasonable proportions, where the gain from such measurements will at least balance the expense.

While the technology for measuring jet streams with new sensing systems makes rapid strides, techniques of analysis and forecasting methods will have to be upgraded. With data pouring in from satellites, radiosondes, and surface observatories, we are struggling to sort, screen, check, digest, disseminate. Obviously, the minds of scores of weather technicians will not be sufficient to control this flood, and to make proper use of it. The intelligence of electronic computers will have to be used. Instead of mapping the jet streams by hand, we will have to feed wind and temperature data into computers for automatic conversion into analyses. Such computer programs have been tested and are in operation, but they still need refinement.

So far, machines have been instructed to "look" only at the large-scale aspects of atmospheric motion and structure. What about the small "wiggles" which we have found in vertical wind profiles? We have to ignore them because they would throw the present "one-track minds" of our computers off their tracks. Eventually, they too will have to be digested. We know that the small-scale turbulence of the atmosphere is especially powerful near jet streams. It may generate the hazardous flying conditions known as clear air turbulence. Strong wind shears, as we find them near the jet stream, will produce strong shearing stresses which, in turn, will cause the smooth flow of the winds to break down into small eddies, measuring a few meters to a few hundred meters in diameter. These eddies may shake up an aircraft very badly, causing injury to passengers, and structural damage and material fatigue to the plane. So far we have been rather inept in forecasting this turbulence, but engineers and scientists are hard at work. They probe the atmosphere and its jet streams with radar and laser beams, hoping that the turbulent eddies may scatter electromagnetic waves which can be detected with apparatus at the ground or in planes, first step in developing a warning system on clear-air turbulence.

All these efforts are still very much in their infancy, but at least we know where the problems lie. If there is anything this century has produced, it is the spirit of teamwork and cooperation among the many disciplines of sciences, and among individual scientists. By defining the problems of research, we already have taken the most important step—we have smoked the enemy out of his cave. He stands in the open now, waiting to be attacked by experimenters, theoreticians, scientists, engineers, and government appropriation committees.

As we go high above the tropopause in the quest of knowledge on stratospheric and mesospheric jet streams, our avenue of approach narrows into a footpath in a jungle of technical difficulties, high costs, and crudities of theory and technique. In time we may have remote sensors, carried by satellites, to record minute quantities of radiation emitted by rare constituents of the atmosphere, such as ozone. Development work on such sensing probes has started, but is yet a long way from perfection. Rocket soundings, still very costly at the present time, may become more economical and efficient. Scientists traveling on a manned orbital laboratory may be equipped with more sophisticated instruments than an automatic, unmanned satellite could handle. They may be able to record data from these lofty regions of the atmosphere and transmit them back to earth.

To try to guess what lies beyond would take us into the realm of science fiction. It would be equally rash to promise the accomplishment of *exact* weather forecasts for weeks and months ahead of time as a result of all these developments. As we have said earlier in this book, it is impossible to measure everything everywhere all the time. Even if we could, who would process the data? What we may look forward to, however, is a steady increase in the statistical reliability of forecasts. Short-period predictions on small-scale weather phenomena, such as tornadoes, squall lines, clear-air turbulence, and so forth, will improve drastically, even though they may still fall short of a perfect score. So will the two- to three-day forecasts of large-scale weather, such as the movement of hurricanes and typhoons, of blizzards and dust storms. Seasonal estimates of weather will become a notch more reliable, too.

Although we still will not be able to claim a perfect score in weather prediction, the foreseeable improvements over the present will be measured in billions of dollars. What is it worth to save a hundred lives from a tornado that is about to strike a crowded shopping center? How do we measure the economic value of families, and homes, and livestock, and machinery, saved in time from onrushing tides driven by a hurricane? Will our increased knowledge of the whims of weather give us the dreamed-of weapon to control at least some of its blind fury? Can we divert typhoons? Can we relieve a drought by gentle rains? Can we avert a leashing hailstorm by softening it up? Even though a Yes to these questions is not for next year's almanac, we are on the way. It will be a long, thorny, frustrating road with many as yet unknown detours. But the challenge is a real one.

APPENDIX

A BASIC REFRESHER IN TRIGONOMETRY

Instead of giving the angles of a right-angled triangle in degrees, we may express them as the ratio of the sides. We do this by using standard nomenclature accepted in trigonometry (Figure 78).

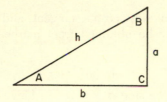

The sine of A is given by the ratio of the sides $\frac{a}{h}$, the cosine of A by $\frac{b}{h}$, where h is the long side, or hypotenuse, b is the base, and a the side of the right-angled triangle. We write

$$\sin A = \frac{a}{h}$$

$$\cos A = \frac{b}{h}$$

The tangent of A is given by $\frac{a}{b}$, the cotangent of A by $\frac{b}{a}$. We write

$$\tan A = \frac{a}{b}$$

$$\cot A = \frac{b}{a}$$

In a similar manner we may state

$$\sin B = \frac{b}{h}$$

$$\cos B = \frac{a}{h}$$

$$\tan B = \frac{b}{a}$$

$$\cot B = \frac{a}{b}$$

The values of sine, cosine, tangent, and cotangent for any angle may be obtained from mathematical tables. For instance

$$\sin 30° = 0.5$$

meaning that the side h is twice as long as side a, thus $\frac{a}{h} = \frac{1}{2}$.

$$\tan 45° = 1$$

meaning that sides a and b are equally long, forming an isosceles triangle.

We may also write, for instance,

$$b = a \cdot \cot A$$

if we wish to compute the length of the base of the triangle with the side and angle A given.

SUGGESTED READING

Science Study Series

Battan, Louis J., *The Nature of Violent Storms*, S 19
—— *Radar Observes the Weather*, S 24
—— *Cloud Physics and Cloud Seeding*, S 29
—— *The Unclean Sky*, S 46
Ohring, George, *Weather on the Planets*, S 47
Page, Robert Morris, *The Origin of Radar*, S 26

Other Books

Dunn, W. L., *Meteorology*. New York: McGraw-Hill (1965), 484 pp.
Petterssen, S., *Introduction to Meteorology*. New York: McGraw-Hill (1958), 327 pp.
Riehl, H., *Introduction to the Atmosphere*. New York: McGraw-Hill (1965), 365 pp.

INDEX

Airlines, jet streams and econo-
mies of, 23–24
Airplanes
clear-air turbulence (CAT)
and, 171–73
fly at constant pressure altitude, 65
upper-air observations by,
41–49, 137
Air speed, true vs. indicated,
19–20
Air trajectories, defined, 81
Altimeter corrections, defined,
85
Altimeters, defined, 65
Anderson, O. A., 2
Anemometers, 25
Angular momentum, 93–103
defined, 94
in "dishpan experiments,"
108
in general circulation, 100–3,
107
of subtropical jet stream,
126–27
tropopause and, 120–21
Angular velocity, defined, 96
Anticyclones (highs), 68, 71,
109, 114
in "dishpan experiments,"
110–11
high-index patterns, 111, 133
subtropical belt of, 106, 123,
135
seasonal migration of, 167
tropical easterly jet stream
and, 136–37
surface vs. upper-air, 118–21
Anticyclonic wind shears, 53,
145

Arctic front jet stream, 124
Astronomy, use of Doppler effect in, 47–48
Atmosphere
angular momentum of, 98
density of, 117–18
energy cycle in, 124–30
general circulation of, 99–
103
pressure forces in, 103–5
(see also Meridional circulation system)
structure of, 9–14
Atomic contamination. See Ra-
dioactive-contaminated
air
Atomic Energy Commission,
monitoring stations of,
159
Azimuth angle, defined, 28
Azores high, 135

B-29 bombers, in discovery of
jet stream, 4–7
Backing, 50, 88
Balloons
manned flights in, 2–3
See also Upper air observa-
tions
Bermuda high, 135
Berson, A., 2, 140
Berson westerlies, 140
Bjerknes, Vilhelm, 148
Blizzards, 109, 130, 151

CAT (clear-air turbulence),
171–73
warning system for, 176
C.G.S. system, 17